laboratory manual
to accompany
physics
for the
life
sciences

laboratory manual to accompany physics for the life sciences

Alan H. Cromer
Professor of Physics
Notheastern University

Robert I. Boughton
Assistant Professor of Physics
Northeastern University

New York St. Louis San Francisco Düsseldorf Johannesberg Kuala Lumpur
London Mexico Montreal New Delhi Panama Paris São Paulo Singapore
Sydney Tokyo Toronto

**McGraw-Hill
Book Company**

Laboratory Manual to accompany Cromer:
PHYSICS FOR THE LIFE SCIENCES
Copyright © 1974 by McGraw-Hill, Inc. All rights reserved.
Printed in the United States of America. No part of this publication
may be reproduced, stored in a retrieval system, or transmitted, in any
form or by any means, electronic, mechanical, photocopying, recording, or
otherwise, without the prior written permission of the publisher.

07-014432-X

1 2 3 4 5 6 7 8 9 0 V H V H 7 9 8 7 6 5 4 3

This book was set in Press Roman by Publications Development Corporation.
The editor was Jack L. Farnsworth; the designer was Publications Development
Corporation; and the production supervisor was Bill Greenwood.
Von Hoffman Press, Inc. was printer and binder.

CONTENTS

Preface		vii
Introduction		1
	mechanics	6
Experiment 1	Measurement	7
	Report Sheets	13
Experiment 2	Force	15
	Report Sheets	21
Experiment 3	Torque	23
	Report Sheets	27
Experiment 4	Acceleration	29
	Report Sheets	35
Experiment 5	Work and Energy	41
	Report Sheets	45
Experiment 6	Mechanical Equivalent of Heat	49
	Report Sheets	55
	properties of matter	58
Experiment 7	Pressure in Fluids	59
	Report Sheets	65
Experiment 8	Fluid Flow	69
	Report Sheets	73
Experiment 9	Boyle's Law	77
	Report Sheets	79
Experiment 10	Temperature	81
	Report Sheets	85
Experiment 11	Vapor Pressure	87
	Report Sheets	91
Experiment 12	Surface Tension	93
	Report Sheets	99
	wave phenomena	102
Experiment 13	Standing Waves	103
	Report Sheets	109
Experiment 14	Interference and Diffraction	113
	Report Sheets	117
Experiment 15	Reflection and Refraction	119
	Report Sheets	125
Experiment 16	Lenses	129
	Report Sheets	133
	electricity and magnetism	137
Experiment 17	The Electric Field	138
	Report Sheets	143

Experiment 18	Ohm's Law	145
	Report Sheets	149
Experiment 19	Kirchhoff's Laws	153
	Report Sheets	157
Experiment 20	The Oscilloscope	159
	Report Sheets	165
Experiment 21	The Magnetic Field	169
	Report Sheets	175

modern physics 178

Experiment 22	Atomic Spectra	179
	Report Sheets	183
Experiment 23	Photoelectric Effect	187
	Report Sheets	191
Experiment 24	Nuclear Radiation	193
	Report Sheets	197
Experiment 25	Nuclear Decay	201
	Report Sheets	205
Appendices	Table of Trigonometric Functions	209
	Table of Square Roots	210
	Index of Equipment	211
	Extra Graph Paper	213

PREFACE

The experiments in this manual are designed to accompany the text *Physics for the Life Sciences* by Alan Cromer. Each experiment relates directly to a significant topic discussed in the text, and each chapter (with the exceptions of Chapters 6 and 11) is represented by at least one experiment. Enough introductory explanation is included in each experiment to enable a student to complete the assignment even if he has not yet studied the material in class. All experiments are cross referenced to the text.

Every instructor probably has his own list of the functions of an introductory physics laboratory. We believe that for life science students, these functions are:

1 To give students direct experience with physical phenomena.
2 To relate the abstract symbols used in physics formulas to their concrete correlates.
3 To demonstrate some of the important laws of physics in order for the students to better understand them.
4 To teach students to analyze numerical data in a systematic way.
5 To teach students the use of standard physical instruments.

An introductory physics laboratory must be highly structured if an inexperienced student is to successfully complete in a few hours a measurement that may have originally taken years to make. For this reason, this manual gives very explicit directions for each experiment. It is expected that in most cases a student can begin work without further direction, although the instructor will often have to help the student as the experiment progresses.

Report sheets are included after each experiment to enable a student to record and analyze his data in a convenient form. These sheets are to be torn out of the manual and handed to the instructor for evaluation. The student should keep his completed evaluated reports in a loose-leaf notebook.

Alan H. Cromer
Robert I. Boughton

INTRODUCTION

The primary purpose of a physics laboratory course is to give you direct experience with the physical phenomena studied in the classroom. In addition, a physics laboratory teaches techniques of measurement and analysis that are valuable in any laboratory situation. It is hoped that this laboratory course will develop your ability to perform accurate laboratory work and to judge the reliability of work reported in the scientific literature.

Before coming to lab read the experiment you are scheduled to perform. It is not necessary that you understand the experimental procedure in detail, but you should have a general idea of the physical principles involved and the experimental approach to be followed. If you have not yet studied the relevant physics in class, read the references cited in the experiment.

Try to complete both the experiment and the report during the laboratory period. If you can't, at least be sure you understand the analysis required before leaving the lab.

The following information will be useful in preparing your report.

RECORDING DATA

Record the result of every measurement directly on the report sheet accompanying the experiment.† The report sheet is the original record of your experiment, and so data should not be transfered to it from scratch paper. Do not erase or obliterate a mistake. Cross it out with a thin line and write the correction above it. Your objective is to maintain a full record of your experiment, including mistakes, rather than to prepare a polished report.

ANALYSIS

After the data is recorded, make the calculations and plot the graphs indicated in the manual. The arithemtic should be done on a slide rule or electronic calculator if possible; long hand calculations should be done on scratch paper and only the results entered on the report sheet.

QUESTIONS

The questions on the report sheet should be answered on a separate sheet of paper and handed to your instructor with your completed report sheet.

GRAPHS

In most experiments you are asked to plot one or more graphs. For example, suppose you have measured the distance d traveled in time t by a mass accelerating down an inclined plane. The first two columns of Table 1 show how the data might appear on your report sheet.

A graph of distance against time is prepared as follows.

---

† Your instructor may ask you to record your data in a spiral-bound quadrille-ruled notebook. In this case, use the report sheet as a model of how to arrange entries in the notebook.

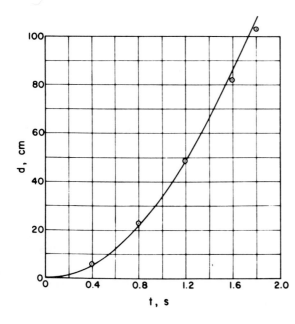

FIGURE 1 Graph of distance (d) against time (t) for an accelerating cart.

1 Label the vertical and horizontal axes with both the quantity it represents and the unit in which the quantity is measured. For example, if the distance traveled by the cart is measured in centimeters, the vertical axis is labeled d, cm, as shown in Fig. 1. Likewise, if time is measured in seconds, the horizontal axis is labeled t, s. Failure to label axes properly is one of the most common mistakes made by students.

2 Choose a convenient scale for each axis. A division of the graph may represent only a decimal multiple of 1, 2, or 5, e.g., 0.1, 2, 20, 0.05, 500. Any other choice, e.g., 0.3, makes the plotting of data difficult. Furthermore, the scale must allow all the data to be plotted over as much of the graph as possible. For example, since the graph in Fig. 1 has 10 horizontal divisions and t goes from 0 to 1.8 s, each division must be 0.2 s, so the horizontal scale goes from 0 to 2 s. The graph also has ten vertical divisions, but since d goes from 0 to 103.2 cm, each division may be taken to be either 10 or 20. The first choice requires you to plot the last point off the graph (Fig. 1), whereas the second choice would use only half of the vertical scale. In this example the first choice is preferred, but, if the last point were at 125 cm, the second choice would probably be better.

TABLE 1 EXAMPLE OF DATA OF A CHART ACCELERATING DOWN AN INCLINED PLANE.

Distance (cm)	Time (s)	t^2 (s^2)
0	0	0
6.0	0.4	0.16
22.8	0.8	0.64
~~28.2~~		
48.6	1.2	1.4
82.0	1.6	~~2.5~~ 2.6
103.2	1.8	3.2

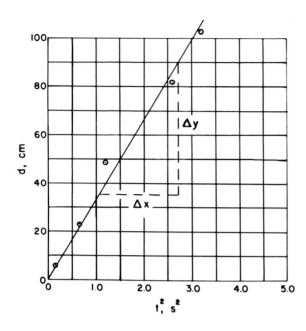

FIGURE 2 Graph of distance (d) against time squared (t^2) for an accelerating cart.

3 Once the scale is chosen, label the axes and carefully plot each point on the graph. Circle the points to distinguish them from accidental marks on the paper. If more than one set of data is plotted on the same graph, use other symbols, e.g., X, Δ, to distinguish the points.

4 Draw a simple smooth curve through the points. Such a curve will not pass through all the points, but it should pass as close as possible to each point, with about half the points on each side of the curve. A French curve is useful for drawing curves. Do not connect the points by straight lines.

STRAIGHT-LINE PLOTS

It is difficult to judge whether an experimental curve agrees with a formula, so in most experiments the variables are rewritten to produce a straight line relation. For example, the formula for the distance traveled by a cart accelerating down an inclined plane is

$$d = \tfrac{1}{2} g \sin \theta \, t^2 \qquad \qquad 1$$

where g is the acceleration of gravity and θ is the inclination of the plane. A plot of d against t is a parabola, but it is not clear from inspection whether the curve in Fig. 1 is consistent with Eq. 1. To test the equation, we calculate t^2 from the data in Table 1 and plot d against t^2 (Fig. 2). Note how the horizontal axis is labeled in this case. With a ruler, a straight line is drawn through the origin of the graph as close as possible to all the points. The degree to which the data are consistent with the equation is shown by how close the points are to the line.

The general equation of a straight line is

$$y = a + sx$$

where a is the value of y when $x = 0$ (the y intercept) and s is the *slope*. The slope is the ratio $\Delta y/\Delta x$ of the vertical rise Δy to the corresponding horizontal displacement Δx of the straight line. From Fig. 2 the slope is found to be

$$s = \frac{\Delta y}{\Delta x} = \frac{56 \text{ cm}}{1.7 \text{ cm}} = 33 \text{ cm/s}^2$$

From Eqs. 1 and 2 we see that the slope is expected to be

$$s = \tfrac{1}{2} g \sin \theta$$

Suppose θ is known to be 4°. Then the experimental value of g found from these measurements is

$$g = \frac{s}{\tfrac{1}{2} \sin \theta} = \frac{33 \text{ cm/s}^2}{0.5 \times 0.070} = 943 \text{ cm/s}^2$$

which is to be compared to the accepted value of 980 cm/s^2

EXPERIMENTAL ERROR

Is the experimental value we obtained for g consistent with the accepted value? A complete answer to this question requires a thorough analysis of all the potential sources of error in the measurement, which is beyond the scope of this course. However, you should be generally aware of how errors in a measurement affect the degree of agreement to be expected.

The *relative error* R_M of a measurement M is the ratio

$$R_M = \frac{|M_{\text{exp}} - M_{\text{th}}|}{M_{\text{th}}}$$

where $|M_{\text{exp}} - M_{\text{th}}|$ is the magnitude of the difference between the experimental and theoretical values of M. R_M is usually expressed as a percentage. In the present example, the relative error in the measurement of g is

$$R_g = \frac{|g_{\text{exp}} - g_{\text{th}}|}{g_{\text{th}}} = \frac{|943 - 980|}{980} = \frac{37}{980} = 0.038 = 3.8\%$$

Whether this error indicates a discrepancy between theory and experiment depends on the error expected from the known uncertainties in the experiment.

From Table 1 we see that t was measured only to the nearest tenth second, so the uncertainty in t is ±0.05 s. Thus, the relative uncertainty in the 1.8-s measurement is

$$R_t = \frac{0.05 \text{ s}}{1.8 \text{ s}} = 0.028 = 2.8\%$$

The uncertainty in $t^2 = (1.8 \text{ s})^2 = 3.24 \text{ s}^2$ is even greater, for if the true value of t can be as small as 1.75 s, the true value of t^2 can be as small as $(1.75 \text{ s})^2 = 3.06$ s, and so the relative uncertainty in t^2 is

$$R_{t^2} = \frac{|3.06 - 3.24|}{3.24} = 0.056 = 5.6\%$$

The expected relative error in g depends on the relative uncertainties of all the measurements that go into the determination of g. Although the details of this are beyond the scope of this course, it is clear that if t^2 is not known to be better than 5.6%, it is not unreasonable to obtain a 3.8-percent error in g. In this case, we say that the measurement is consistent with theory to within the known uncertainties of the measurement. This is all that can ever be said of any measurement.

The expected relative error for most of the experiments in this manual is between 3 and 10 percent. The main objective of these experiments is to give you direct experience with physical phenomena, rather than to obtain great accuracy.

mechanics

EXPERIMENT 1   MEASUREMENT

GOALS

1 To learn the metric units of length, mass, and volume.

2 To learn to measure length and mass with a ruler, vernier caliper, micrometer, and balance.

3 To understand the concepts of mass, volume, and density.

4 To learn the graphical method of data analysis.

EQUIPMENT

Four metal cylinders   Micrometer
Ruler   100-ml graduated cylinder
Vernier Caliper   Triple-beam balance

INTRODUCTION

Experiments in physics generally involve the measurement of a physical quantity, and physical data consist of the numbers that are the results of these measurements. (This is in contrast to more qualitative sciences, such as biology, in which the data often consist of drawings and descriptions, rather than numbers.) Furthermore, almost all measurements ultimately involve determining the position of a reference mark on a graduated scale.

A ruler is perhaps the most familar example of a scale, but the English ruler, which is divided into *inches* (in.) and nondecimal fractions of an inch (e.g., eighths, sixteenths), is unsuitable for scientific work. In scientific work length is measured with a metric ruler, which is divided into *centimeters* (cm) and tenths of centimeters, i.e., millimeters (mm). The advantage of the metric scale over the English scale is illustrated in Fig. 1.1, which shows the length of a cylinder being measured by both scales. On the English scale the length of the cylinder is read as 2 in., plus 2 eighth-in., plus ¼ eighth-in. To express this as a single number, it is necessary to first write $2 \times \frac{1}{8} + \frac{1}{4} \times \frac{1}{8}$ as a decimal. On the other hand, the same length on the metric scale is read as 5 cm, plus 7 mm, plus 0.9 mm, which can be immediately written as 5.79 cm, because 1 mm = 0.1 cm.

The last digit in 5.79 is found by estimating the fractional distance of the end of the cylinder from the 5.7 line to the 5.8 line. Although not as accurately determined as the first two digits, the last digit is significant and should be recorded. It would be incorrect, however, to record the number as 5.790 cm, because the last digit (0 in this case) is not significant.

Table 1.1 gives the most commonly used metric units of length, volume, and mass, and the conversions between them. These units will be used throughout your study of physics, so be sure that you know them.

REFERENCE Cromer: *Physics for the Life Sciences*, Sec. 1.2, 1.3, Appendix VIII.

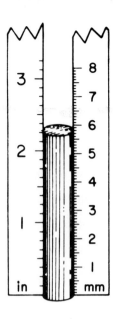

FIGURE 1.1 The length of a cylinder measured with an English ruler and a metric ruler.

TABLE 1.1 COMMON METRIC UNITS

Physical quantity	Units of measurement	Abbreviation	Conversion relations
Length	meter	m	1 m = 100 cm = 1000 mm
	centimeter	cm	1 cm = 0.01 m = 10 mm
	millimeter	mm	1 mm = 0.1 cm = 0.001 m
Volume	liter	l	1 l = 1000 ml
	milliliter	ml	1 ml = 0.001 l
	cubic centimeter	cm^3	1 cm^3 = 1 ml
Mass	kilogram	kg	1 kg = 1000 g
	gram	g	1 g = 0.001 kg

PROCEDURE

You will be supplied with four metal cylinders of different sizes, labeled A, B, C, and D. All four cylinders are made of the same metal, either aluminum, brass, or steel.

Measure the lengths L and the diameters d of each cylinder with a ruler. Each measurement can be made to an accuracy of about 0.02 cm by estimating the fractional part of a millimeter. Be sure to record your measurements to the correct number of significant figures.

Repeat these measurements using a vernier caliper to measure the lengths and a micrometer to measure the diameters. A *vernier caliper* (Fig. 1.2) is a device that enables the ends of an object to be precisely aligned with a ruler. It also has a *vernier scale* that helps to estimate the fractional parts of a division. The measured distance is indicated by the position on the main scale of the first vernier line (0). In the example shown in Fig. 1.3, the vernier indicates a distance of 3.2+ cm. The fractional distance from 3.2 cm to 3.3 cm is represented by the number of divisions between the first vernier line and the vernier line that most closely coincides with a line on the main scale. In Fig. 1.3 for instance, the fourth vernier line coincides with a line on the main scale. Each of the 3 divisions between the first and the fourth vernier lines corresponds to 1/10 of the smallest subdivision on the main scale, so the full reading of the caliper in Fig. 1.3 is 3.23 cm.

FIGURE 1.2 Vernier caliper
[*Central Scientific Co.*]

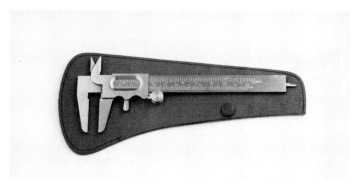

FIGURE 1.3 Use of a vernier scale.

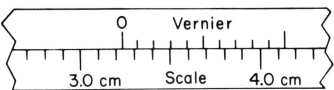

A *micrometer* (Fig. 1.4) is a device that can measure small distances to an accuracy of better than 0.001 cm. Before you make a measurement with a micrometer, you should first check its alignment by closing its jaws with nothing between them. (To avoid damaging a micrometer by closing it too tightly, always turn the shaft by the small rachet on the end.) If the reference line on the body of the micrometer is not opposite the zero mark on the circular scale, the micrometer is misaligned. In that event all your subsequent measurements must be corrected by adding (or subtracting) the number of divisions that the zero mark is behind (or ahead) of the reference line when the micrometer is closed.

To measure the diameter of a cylinder close the jaws of the micrometer on the cylinder until the rachet starts to slip. The linear scale on the shaft of the micrometer is graduated in millimeters (0.1 cm). The fractional part of a millimeter is read on the circular scale. The micrometer shaft must be turned through two complete revolutions to advance the shaft 0.1 cm, and the circular scale on the shaft is divided into 50 divisions, so each division on the circular scale represents 0.001 cm. By estimating the fractional distance between **these divisions**, the micrometer can be read to 0.0001 cm. It is important

FIGURE 1.4 Micrometer
[*Central Scientific Co.*]

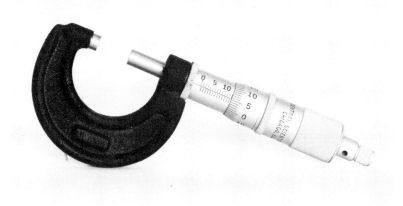

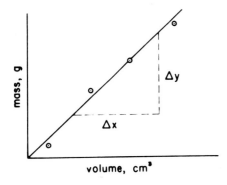

FIGURE 1.5 Graph of mass against volume. The straight line is drawn so that it passes through the origin and as close as possible to the data points.

to remember that 0.050 cm must be added to the reading on the circular scale whenever the shaft is more than halfway between two 0.1-cm divisions on the linerar scale.

Measure the volume V of the two largest cylinders by determining the volume of water they displace when put into a 100-ml graduated cylinder half filled with water. (Be careful to tip the graduated cylinder so that the metal cylinder slides gently into it.)

Measure the mass m of each cylinder on a triple-beam balance.

ANALYSIS

The *volume* V of a cylinder is related to its length L and its diameter d by

$$V = \tfrac{1}{4}\pi L d^2$$

where $\pi = 3.14$. Calculate the volume of each cylinder from your measurements of L and d. Keep only significant figures in your final results. Compare the calculated values of V with the values obtained directly by displacing water in a graduated cylinder.

The *density* ρ of an object is the ratio of its mass to its volume

$$\rho = \frac{m}{V} \qquad 1$$

The proper metric units of density are either grams per cubic centimeter (g/cm³) or kilograms per cubic meter (kg/m³). Calculate the density of each cylinder from your data.

The density of a homogeneous substance is an intrinsic property of the substance, independent of its mass. Therefore, except for errors of measurement, you should obtain the same value of ρ for each cylinder. Your four determinations of ρ will not all be the same, however, because of the limited accuracy of each measurement.

Use the following two methods to obtain a "best value" of the density of your cylinders from your data:

Method 1 Average the four values of ρ.

Method 2 Plot a graph of the mass m of each cylinder against its volume V. From Eq. 1 we have

$$m = \rho V$$

which means that the plot of m against V should be a straight line that passes through the origin. Draw a straight line through your points. If your points do not lie on a straight line draw a line that best fits your data, as shown in Fig. 1.5. The "best value" of ρ is given by the *slope s* of the line. The slope is the ratio of the vertical displacement Δy to the horizontal displacement Δx

$$s = \frac{\Delta y}{\Delta x}$$

Calculate the slope s of your line. Compare this value of to the value obtained by Method 1.

Compare your value of ρ with the density given in a table of densities. (See Cromer: *Physics for the Life Sciences,* Table 7.2, or *Handbook of Chemistry and Physics*, Chemical Rubber Co., Cleveland.)

Name _____

Date _____

REPORT SHEET

EXPERIMENT 1   MEASUREMENT

DATA   Type of metal cylinder used (aluminum, brass, or steel) _____

Cylinder	Ruler		Caliper and micrometer		Displacement	Scale
	L (cm)	d (cm)	L (cm)	d (cm)	V (cm^3)	m (g)
A						
B						
C						
D						

ANALYSIS

1 Calculate the volume, $V = \frac{1}{4}\pi L d^2$, and the density, $\rho = m/V$, of each cylinder.

Cylinder	Calculation	Scale	ρ (g/cm^3)
	V (cm^3)	m (g)	
A			
B			
C			
D			

Average value of ρ = _____

2 Plot the mass m against V, and determine ρ from the slope of the straight line drawn through the data points.

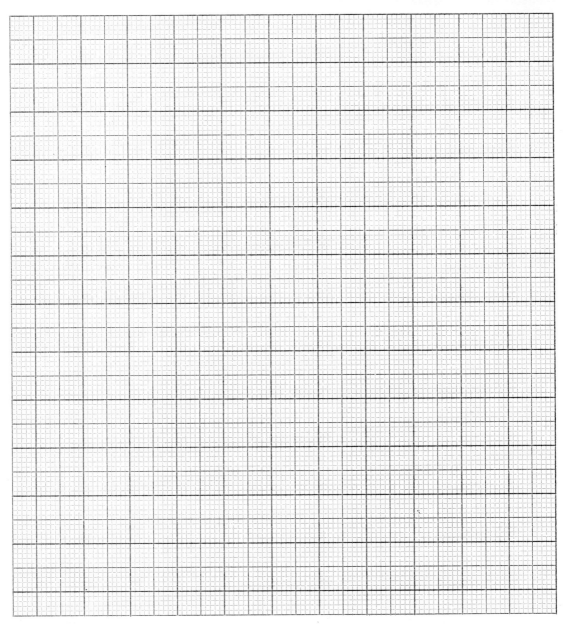

Slope, $s = \rho = $ _____

QUESTIONS

1. Which method of determining the volume of a cylinder is more accurate? Why can't the displacement method be used to measure the volumes of the smaller cylinders?

2. State the advantages and disadvantages of the two methods used to obtain the best value of the density.

3. Estimate the volume of your partner's brain from measurements of his skull made with a ruler. If the density of the brain is 1 g/cm^3, what is the mass of your partner's brain?

EXPERIMENT 2 FORCE

GOALS

1 To learn to add forces vectorially.

2 To learn how Newton's first law is used to find an unknown force.

3 To become familiar with some common arrangements of pulleys, cords, and weights.

EQUIPMENT

Force board (compact type)	Pulleys
Protractor	Cords
Spring scales	Weights

INTRODUCTION

A *force* is the push or pull that one object exerts on another. The essential properties of force are summarized in Newton's three laws of motion. One consequence of Newton's first law is that *if an object is at rest the total force on it is zero.* In this experiment, you will make a quantitative study of this law.

The first law is a statement about an object, and in any given situation the object in question must be recognized. That is, while the law is true for any object, in any particular situation we are always concerned with a particular object. Only when this object is identified and (mentally) isolated from the rest of the world, can we identify the forces that act on it. The total force on the object is the sum of the individual forces acting on it.

A force is characterized by both its magnitude and its direction. (Such quantities, which are characterized by both magnitude and direction, are called *vectors*.) A force is conveniently represented by an arrow that points in the direction of the force. The length of the arrow is chosen to represent the

FIGURE 2.1 An arrow representing the magnitude and direction of a force.

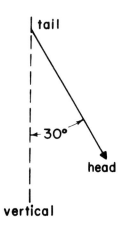

FIGURE 2.2 Vector addition of two forces.

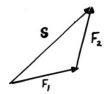

magnitude of the force according to some arbitrary scale. For example, to represent a 4-lb force acting at an angle of 30° to the vertical, a line is first drawn to represent the vertical (Fig. 2.1). Then another line is drawn at 30° to the first line, to represent the direction of the force. Finally, the length of the arrow is drawn to scale. For instance, with the scale 1 lb = 1 cm, the arrow in Fig. 2.1 would be drawn 4 cm long.

The sum **S** of two forces $\mathbf{F}_1$ and $\mathbf{F}_2$ is found by first drawing the forces to scale, with the tail of $\mathbf{F}_2$ coincident with the head of $\mathbf{F}_1$. Then **S** is represented by the arrow drawn from the tail of $\mathbf{F}_1$ to the head of $\mathbf{F}_2$, as shown in Fig. 2.2. All vector quantities are added this way. Like a force, any vector quantity can be represented by an arrow drawn to scale. The sum of three or more forces is found by drawing the forces head to tail, and then drawing **S** from the tail of the first force to the head of the last.

Figure 2.3(*a*) shows that if the sum of two forces is zero, the forces must be equal and opposite. Figure 2.3(*b*) shows that if the sum of three forces is zero, the forces must form a triangle when arranged head to tail so that the distance from the tail of $\mathbf{F}_1$ to the head of $\mathbf{F}_3$ is zero. Of course, if the three forces are parallel to each other, the triangle collapses into a straight line, as shown in Fig. 2.3(*c*).

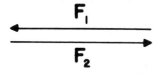

FIGURE 2.3 Examples of forces that add to zero. (a) Two forces that add to zero must always be equal and opposite. (b) Three forces that add to zero must form a closed triangle. (c) In special cases where three forces are parallel, the triangle collapses to a straight line.

a

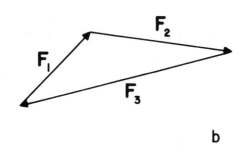

b

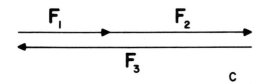

c

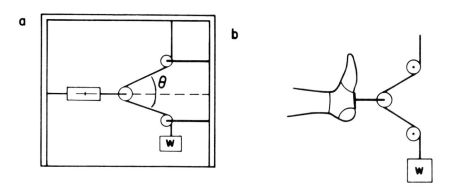

FIGURE 2.6 (a) System of pulleys and weight similar to that used in a leg traction apparatus. (b) A leg in traction.

Set up the arrangement shown in Fig. 2.6(a). This arrangement is similar to one commonly used to apply traction to a patient's leg [Fig. 2.6(b)]. Put a known weight W on the cord and measure the angle θ between the cords. Record after changing W and θ.

Set up some of the arrangements shown in Fig. 2.7. In each case, make a sketch of the arrangement and record the weights, scale readings, and angles.

ANALYSIS

For each arrangement, calculate the tension in the cord containing the scale in terms of W and θ. Compare your result with the scale reading observed.

FIGURE 2.7 Several arrangements of cords, pulleys, and weights.

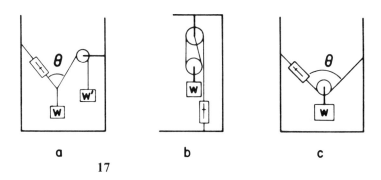

REFERENCE Cromer: *Physics for the Life Sciences,* Chap. 2.

Part I Springs

Springs have the property that the force they exert is proportional to the distance they are stretched. A spring scale is simply a spring attached to a marker that indicates the distance the spring is stretched on a graduated scale. Thus, a spring scale provides a simple way to measure force.

PROCEDURE

The apparatus consists of three spring scales connected to a ring (the object). Chains attached to the other end of each scale can be inserted into slots located around the circumference of a circular force board (Fig. 2.4). In this way, the force exerted on the ring by each scale can be varied independently. The geometry of a particular arrangement is found by tracing the positions of the scales on a piece of paper placed under the scales. Clips on the board hold the paper in place.

FIGURE 2.4   Force board, compact style [*Sargent-Welch Scientific Co.*]

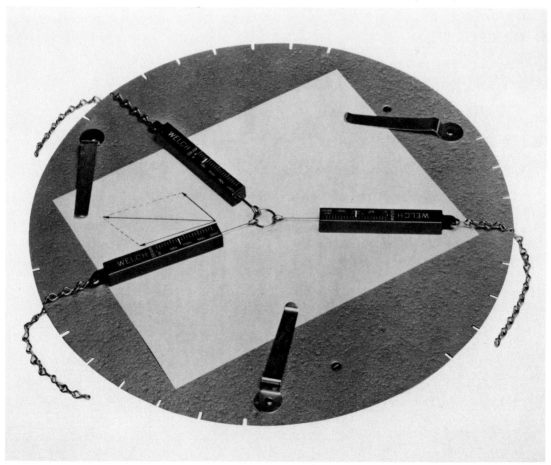

For each of the following arrangements, draw a schematic diagram, and measure and record the magnitudes of the forces applied to the ring and the angles between these forces.

1. Take two of the scales attached to the ring and connect their chains into two slots on opposite sides of the board. Leave the third scale disconnected. Record the magnitudes of the forces and the angle between them. Repeat several times, changing the distance between the two slots each time.

2. Insert the chain of one scale into one slot, and the chains of the other two scales into another slot. Record the results. Repeat using different slots.

3. Insert the three chains into three different slots. Record the results. Repeat after changing the positions and lengths of the chains.

ANALYSIS

In each arrangement the vector sum of the forces applied to the ring should be zero. Calculate the vector sum of the forces measured in each case, and draw an arrow to represent it. The nonzero value of the vector sum is a measure of the error in the experiment.

Part II Cords

A flexible cord exerts a force equal to the tension T of the cord. As long as the cord is not rubbing against (that is, not applying a tangential force to) another surface, the tension in it is the same everywhere. A pulley changes the direction of a cord without changing its tension, because the pulley rotates until the total tangential force applied to it is zero.

PROCEDURE

To demonstrate that the tension is not changed when a cord passes over a pulley, attach a weight W to one end of a cord, pass the cord over a pulley, and connect the other end to a spring scale (Fig. 2.5). Note that the scale reading does not change as you change the angle θ between the two straight sections of the cord. This reading is the tension in the cord. Compare T to the weight W. Note what happens when your partner prevents the pulley from rotating by holding it in a fixed position.

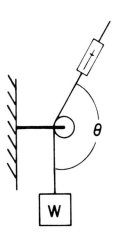

FIGURE 2.5   The scale reading does not depend on the angle θ.

Name_____

Date_____

REPORT SHEET

EXPERIMENT 2   FORCE

Part I   Springs

DATA   Make a sketch below of each arrangement of springs, and on the sketch indicate the reading of each scale and the angles between the scales.

ANALYSIS   For each arrangement of springs, find the sum of the forces applied to the ring. (Use additional paper as needed.)

QUESTIONS

1 (a)   What is the largest force that can be formed by adding an 8-lb force and a 5-lb force? (b)   What is the smallest force that can be formed from these forces? (c)   Show how these two forces can be added to form an 8-lb total force.

2 If only three forces of equal magnitude act on a stationary object, what must be the angles between them?

3 Show that the sum of a 12-lb, a 7-lb, and a 3-lb force cannot be zero, regardless of the angles between them.

Part II   Cords

DATA   Sketch and number each arrangement of pulleys you studied, and record the measurements in the data table.

Arrangement	W	W'	θ	T (scale reading)	T (calculated)

ANALYSIS   For each arrangement of pulleys, calculate the tension in the cord containing the scale and enter on the data table. Compare your result with the observed scale reading.

QUESTIONS

1   Why, when the pulley in Fig. 2.5 is held fixed, does the scale reading change as the angle θ is varied?

2   For what angle θ is the scale reading in Fig. 2.6(a) greatest? For a given weight W, what is the maximum force that can be applied to the leg by the traction device in Fig. 2.6(b)?

3   Explain why a cord with a weight suspended from it [Fig. 2.7 (c)] cannot be stretched perfectly straight.

EXPERIMENT 3   TORQUE

GOALS

1 To learn to calculate the torque applied by a force about a point.

2 To demonstrate that the sum of the torques on an object at rest is zero.

3 To use torques to find the center of gravity of an object.

EQUIPMENT

Weighted meter-stick
   (pivoted at one end)
Cord
Protractor

Spring scale
Weights
Meterstick
Two triple-beam balances
   (with inverted-V supports)

INTRODUCTION

The necessary and sufficient conditions for an object to be in equilibrium are that the total force and the total torque on the object be zero. The first condition is studied in Experiment 2 and the second condition is studied in this experiment.

Torque τ is a measure of the tendency of a force **F** to produce rotation about a specified point O. It is equal to the magnitude F of the force times the perpendicular distance d from O to the line of action of the force. In symbols we have

$$\tau = Fd$$

Figure 3.1 shows how the distance d is defined.

A torque is positive if the force tends to cause counterclockwise rotation about O, and it is negative if the force tends to cause clockwise rotation. At equilibrium the sum of the torques caused by all the forces acting on an object is zero.

Of particular importance is the force of gravity on an object. For the purpose of calculating torque, this force, which is the weight of the object, can be considered to act at a single point called the *center of gravity* of the object. You will learn how to locate the center of gravity of an object in Part II of this experiment.

REFERENCE   Cromer: *Physics for the Life Sciences*, Chap 3.

Part I   Torques on A Rod

PROCEDURE

This part of the experiment consists of three series of measurements of the torques applied to an object in equilibrium. The object is a weighted meterstick that is pivoted at one end.

1 Attach one end of a spring scale to a notch on the meterstick and connect the other end to a cord

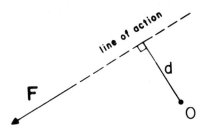

FIGURE 3.1 The distance d from a point O to the line of action of a force.

suspended from the horizontal bar tht is mounted above the stick (Fig. 3.2). Adjust the length and position of the cord until it is vertical and the meterstick is horizontal. Record the scale reading F_s and the perpendicular distance d from the pivot point O to the line of action of the spring force. Move the scale to a different notch on the stick and repeat the procedure, making sure the cord is vertical and the stick is horizontal. Make measurements for at least four different positions of the scale along the meterstick.

2 Suspend a weight W of several hundred grams from the end of the meterstick. Measure the distance D from O to the line of action of W. Make measurements of the scale reading F_s and the distance d from O to the line of action of $\mathbf{F}_s$ for several different positions of the scale. Keep the cord vertical and the stick horizontal for each measurement. (See Figure 3.3.)

3 With the weight W still suspended from the end of the meterstick, change the direction of the force $\mathbf{F}_s$, as shown in Fig. 3.3. Measure the scale reading F_s and the perpendicular distance d from O to the line of action of $\mathbf{F}_s$. Repeat for different values of the angle θ.

ANALYSIS

1 For each position of the scale in the first series of measurements, calculate the torque τ_s applied by the scale about the pivot O.

The forces on the meterstick are its own weight $\mathbf{F}_g$, the spring force $\mathbf{F}_s$, and the force at the pivot $\mathbf{F}_c$. Since $\mathbf{F}_c$ exerts zero torque about the pivot, the sum of the torque τ_s exerted by $\mathbf{F}_s$ and the torque τ_g exerted by $\mathbf{F}_g$ is zero:

$$\tau_s + \tau_g = 0$$

or

$$\tau_g = -\tau_s$$

Since τ_g does not depend on the position of the scale, each measurement should give the same value of τ_s. Average the values of τ_s obtained in the first series of measurements to get your "best value" of τ_g.

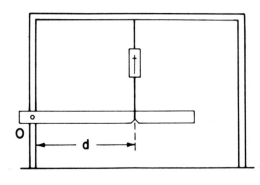

FIGURE 3.2 Pivoted rod supported by a cord attached to a spring scale showing the distance d.

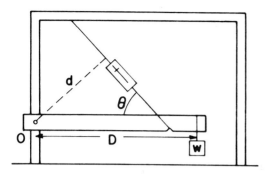

FIGURE 3.3 The direction of the supporting cord is inclined to the vertical.

2 When the weight W is suspended from the stick, the condition for equilibrium becomes

$$\tau_s + \tau_g + \tau_w = 0$$

where τ_w is the torque exerted by W about O. For each position of the cord in the second series of measurements, calculate the torque τ_s applied by the scale. Use the average value of τ_s and the value of τ_g found in 1 to find τ_w. Compare the value of τ_w obtained this way to the value obtained from the formula $\tau_w = WD$.

3 For each angle of the cord in the third series of measurements, calculate τ_s. When the force exerted by the scale is applied at an angle, the torque it must exert to maintain equilibrium does not change, but the perpendicular distance d does. Compare the average value of τ_s found in the third series of measurements to the average value found in the second series.

Part II Locating The Center of Gravity

PROCEDURE

Carefully unbolt the meterstick from the pivot and set the bolt and washers aside where they will not get lost. Set the stick on two scales, as shown in Fig. 3.4. The stick must be placed on inverted-V supports, so that the points of contact between the stick and the scales are well defined. Record the dis-

FIGURE 3.4 Rod supported on two scales to determine the location of its center of gravity.

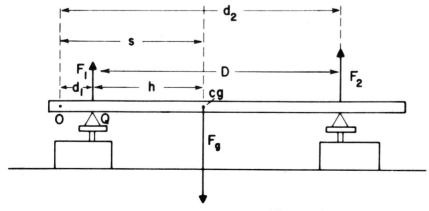

tances d_1 and d_2 from O to the two contact points, and the magnitude F_1 and F_2 of the forces exerted on the stick by the two scales. (Be sure to subtract the weight of the supports from the scale readings.) Repeat after changing the position of the meterstick on the scales.

When you have finished rebolt the stick to the pivot, making sure the washers are in contact with the stick. Have your instructor check that the pivot has been reassembled properly. The stick must move freely in the pivot, so do not tighten the bolt too tightly.

ANALYSIS

The center of gravity of the stick is the point cg at which the total force of gravity $\mathbf{F}_g$ on the stick can be taken to act. Thus, if we take torques about the contact point Q in Fig. 3.4, the condition of equilibrium yields

$$F_g h = F_2 D$$

or

$$h = \frac{F_2 D}{F_g}$$

But the magnitude of $\mathbf{F}_g$ is equal to the sum of the corces F_1 and F_2 that the scales exert on the stick, so we have

$$h = \frac{F_2 D}{F_1 + F_2}$$

From your measurements of d_1, d_2, F_1, and F_2, find the distance s of the center of gravity from the pivot point O.

The torque τ_g that the force of gravity exerts about O is

$$\tau_g = F_g s$$

Compare the value of τ_g found from this formula to the value found in the analysis of the first series of measurements in Part I.

Name _____

Date _____

REPORT SHEET

EXPERIMENT 3   TORQUE

Part I   Torques on a Rod

DATA

Series 1		
F_s	d	τ_s
Average τ_s = _____		

Series 2		
F_s	d	τ_s
Average τ_s = _____		

Series 3		
F_s	d	τ_s
Average τ_s = _____		

Weight, W = _____

Distance, D = _____

ANALYSIS   Calculate the torque τ_s for each series of measurements and enter in data table. Complete the following calculations.

1 τ_g = average τ_s = _____

2 τ_w (calculated from τ_g and τ_s) = _____

   τ_w (calculated from W and D) = _____

3   Average τ_s (Series 2) = _____

   Average τ_s (Series 3) = _____

QUESTIONS

1   Do τ_s and τ_w have the same sign? Explain.

2   Why is the torque exerted about O by $\mathbf{F}_c$ zero?

3   Under what conditions could the pivot exert a nonzero torque about O? Would this be a random or a systematic error in the experiment?

Part II Locating the Center of Gravity

DATA

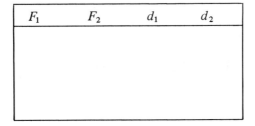

F_1	F_2	d_1	d_2

ANALYSIS

The distance s from O to the center of gravity = _____

The torque τ_g due to the force of gravity = $F_g s$ = _____

The torque τ_g found indirectly in Part I = _____

QUESTIONS

1 Why is F_g equal to $F_1 + F_2$?

2 Describe another method for locating the center of gravity of the meterstick.

3 Show how the mass of your forearm can be determined, if the location of its center of gravity is known.

EXPERIMENT 4   ACCELERATION

GOALS

1 To observe and study the relations between time, distance, and speed for an object moving with constant linear acceleration.

2 To measure the acceleration of gravity.

3 To study the motion of two masses connected by a flexible cord.

EQUIPMENT

Linear air track   Spark-sensitive paper
Spark timer   Pulley and weights

INTRODUCTION

An object moving with constant speed v moves the distance

$$d = vt \qquad\qquad 1$$

in the time t. A plot of d against t is thus a straight line with a slope equal to the speed of the object, as shown in Fig. 4.1. This is equivalent to saying that the speed of the object is the ratio

$$v = \frac{\Delta x}{\Delta t} \qquad\qquad 2$$

where Δx is the distance the object moves in the time interval Δt.

At time t, the speed of an object moving with constant linear acceleration a is

$$v = v_0 + at \qquad\qquad 3$$

and the distance d the object moved is

$$d = v_0 t + \tfrac{1}{2} a t^2 \qquad\qquad 4$$

where v_0 is the speed of the object at time $t = 0$. A plot of d against t is a parabola, as shown in Fig. 4.2. Since the speed of the object is continuously changing, it is not in general given by Eq. 2. However, Eq. 2 can be used to approximate v during short time intervals. For example, during the interval $\Delta t = t_2 - t_1$ the object moved the distance $\Delta x = d_2 - d_1$ shown in Fig. 4.2, so the average speed of the object during this interval is $\Delta x / \Delta t$. This is equal to the constant speed v of an object represented by the straight line that passes through points A and B in Fig. 4.2. In so far as this straight line is a good approximation to the curve between the points A and B, the speed of the accelerating object is constant enough in this interval to be approximated by Eq. 2.

Near the surface of the earth, all objects fall vertically with the constant linear acceleration $g = 980 \text{ cm/s}^2 = 9.80 \text{ m/s}^2$. Because this acceleration is so large, objects fall too fast for the details of their motion to be easily observed. In this experiment, two special mechanical systems will be used that effectively scale down the acceleration of gravity so that accelerated motion can be more easily studied.

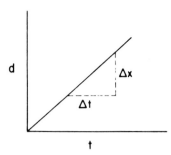

FIGURE 4.1 Plot of distance against time for an object moving at a constant speed.

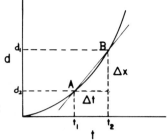

FIGURE 4.2 Plot of distance against time for an object moving with constant linear acceleration.

INCLINED PLANE Figure 4.3 shows an object of mass m sitting on an inclined plane. If the plane is frictionless, the only forces on the object are the contact force $\mathbf{F}_c$, which is perpendicular to the incline, and the force of gravity $\mathbf{F}_g$, which is perpendicular to the horizontal. The only force parallel to the incline is the component of $\mathbf{F}_g$ parallel to the incline. The magnitude of this component is

$$F_x = F_g \sin \theta = mg \sin \theta \qquad 5$$

or

$$a = g \sin \theta$$

By making the angle of the incline very small, we can make the acceleration as small as we want. Note, however, that this is possible only if the incline is completely frictionless, since otherwise the force of friction would prevent the object from sliding when the angle θ was sufficiently small.

PULLEY AND PLANE Figure 4.4 shows an object of mass m sitting on a frictionless horizontal plane. The object is connected by means of a cord and pulley arrangement to a vertically suspended object of mass m'. The tension T of the cord is the only horizontal force on m, so the acceleration a of m is given by

$$T = ma$$

The forces on m' are its weight $m'g$ and the upward force T exerted by the cord. Since m' has the same acceleration as m, we have from Newton's second law

$$m'g - T = m'a$$

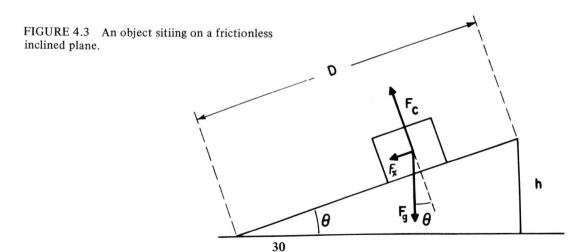

FIGURE 4.3 An object sitiing on a frictionless inclined plane.

30

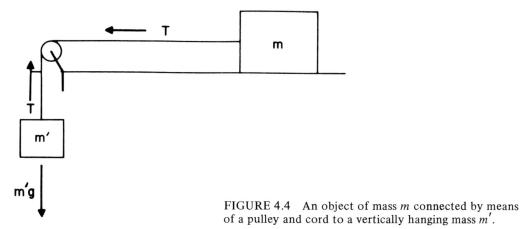

FIGURE 4.4 An object of mass m connected by means of a pulley and cord to a vertically hanging mass m'.

When we add these last two equations the unknown quantity T is eliminated and so we get

$$m'g = ma + m'a$$

or

$$a = \frac{m'}{m + m'} g \qquad\qquad 6$$

Thus, the acceleration of the system can be made as small as we want by making m' sufficiently small. Again, this procedure will work only if the plane and pulley are completely frictionless, so that a very small mass m' is sufficient to accelerate the system.

REFERENCE Cromer: *Physics for the Life Sciences*, Chap. 4.

Part I Inclined Plane

PROCEDURE

The apparatus is a linear air track (Fig. 4.5), which is a track on which small objects, called gliders, slide without friction on a cushion of air. The air is blown into the interior of the track by a blower or a compressed air supply, and it flows out through hundreds of tiny holes on the upper surface of the track. The steady flow of air through these holes supports the glider and keeps it from touching the track, so that there is no frictional force between the glider and the track. There is still a minute drag on the glider due to air resistance but this is negligible at the slow speeds used in air-track experiments.

Place a glider on your track and turn on the air supply. Unless the track is perfectly level the glider will accelerate toward one end. Adjust the leveling screw at the end of the track until the track is so level that the glider moves very little when placed anywhere on the track.

Measure the distance D between the legs of the track, and place a metal block of known height h under the leg beneath the leveling screw. Now when the glider is released it will accelerate down the incline with the acceleration given by Eq. 5, where $\sin \theta = h/D$. In this experiment you will measure a, and thus determine g from the known values of D and h.

The time and position of the glider as it slides down the track are recorded using a *spark timer*.

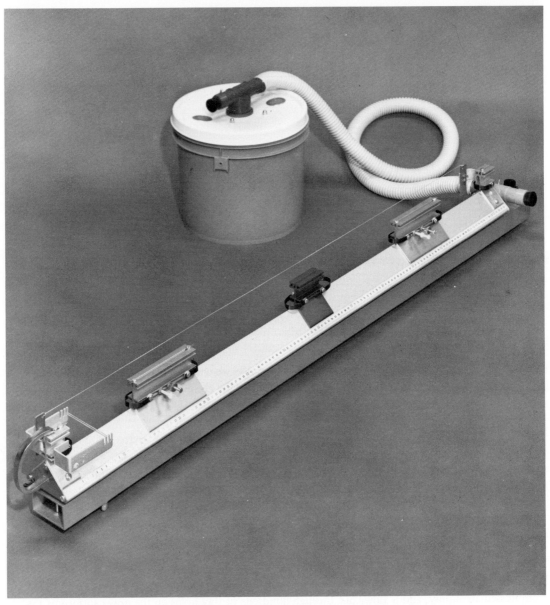

FIGURE 4.5 Linear air track and blower [*Ealing Scientific Co.*]

This is a device that produces a high electrical potential (voltage) at regularly spaced intervals of time. One terminal of the spark timer is connected to a guide wire that runs parallel to the track, and the other terminal is connected to the body of the track (Fig. 4.6). A jump wire attached to the glider is bent so that one end is about a millimeter or so from the guide wire and the other end is about a millimeter or so from the ridge that runs along the side of the track. It is important that the jump wire be as close to the guide wire and ridge as possible without actually touching them. When everything is connected, turn on the spark timer. *Be careful not to touch the track while the spark timer is on, since the high voltage involved can give you an unpleasant shock.* The spark timer produces a steady

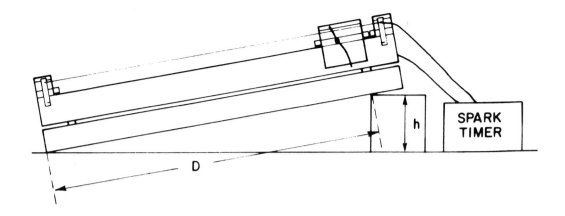

FIGURE 4.6 Connection of a spark timer to an air track.

series of sparks that jump between the guide wire and the top of the jump wire, and between the bottom of the jump wire and the body of the air track. Change the frequency control on the spark timer and observe how the frequency of the sparks changes. These sparks will be used to mark the time and position of the glider on a strip of paper.

To take a measurement, first, tape a strip of fresh spark-sensitive paper along the ridge on the side of the track. Second, tie the glider at the upper end of the track by a short piece of thread. Third, turn on the air supply and the spark timer. Set the frequency of the timer to 10 sparks/s, so that the time interval between sparks is $\Delta t = 1/10$ s. As long as the glider is restrained by the thread, the sparks will all fall within a small area on the paper. The radius of the spot formed by these sparks is the intrinsic uncertainty of the position measurement in this experiment. Finally, release the glider by burning the thread with a match. Be sure to turn off the spark timer just before the glider reaches the bottom of the track, otherwise additional marks will be made when the glider bounces back up the track.

Remove the spark-sensitive paper and examine it to be sure you have a series of clearly distinguishable marks. Record your name, the date, the value of h, and the spark frequency on one end of the tape. The tape must be preserved as it is part of the experimental record.

Take a second measurement with the track inclined at a steeper angle.

ANALYSIS

To analyze a spark record, tape the spark-sensitive paper to a flat surface so that it will not move around as you work on it. Locate the first mark on the paper that is clearly separated from the large spot that was made while the glider was restrained by the thread. Label this mark 0, and number the successive marks on the paper 1, 2, 3, ..., n. Measure the distance Δx_n between adjacent marks. According to Eq. 2, the speed of the glider between two marks is

$$v_n = \frac{\Delta x_n}{\Delta t}$$

where $\Delta t = 1/30$ s is the time interval between the marks on the paper.

Let the time t at position 0 be zero, so that the time t_n at the nth position is

$$t_n = n\Delta t$$

Plot v_n against t_n and draw a straight line through the points. According to Eq. 3, the acceleration a of the glider is equal to the slope of this line. Determine a in this way for both inclinations of the track.

From Eq. 5, the acceleration of gravity is related to a by

$$g = \frac{a}{\sin \theta} = \frac{a}{h/D}$$

Calculate g separately for both values of h, and average the two values of g obtained. Compare the average value of g with the expected value of 980 cm/s^2.

Part II Pulley and Plane

PROCEDURE

Level the air track and tie a glider to one end of the track by a short thread. By means of a cord, connect the glider to a mass m' hanging from a pulley at the other end of the track, as in Fig. 4.4. Be sure that m' is less than one-tenth the mass of the glider.

REMARK Equation 6 is exact only if the pulley is massless. Some air tracks have a curved neck at one end which acts as a massless pulley because it has no moving parts. With these tracks the glider is connected to m' with a piece of plastic tape. Air passing through the holes in the neck enables the tape to slide over it without friction. If you use a regular pulley, your instructor may give you a corrected formula to use in place of Eq. 6.

To make a measurement, proceed as in Part I. Attach fresh spark-sensitive paper, turn on the timer, and burn the restraining thread.

Repeat using a different mass m'.

Measure the mass m of the glider and the two hanging masses m'.

ANALYSIS

Use the method described in Part I to analyze each spark record and find the acceleration a. For each value of a, determine g from Eq. 6. Compare the average value of g to the expected value.

Name _____

Date _____

REPORT SHEET

EXPERIMENT 4   ACCELERATION

Part I   Inclined Plane

DATA

Distance D = _____

Time interval Δt = _____

First Series		
Height, h = _____		
n	Δx_n	v_n
0	—	—
1		
2		
3		
4		
5		
6		
7		
8		
9		
10		
11		
12		
13		
14		
15		
16		
17		
18		
19		
20		
21		

Second Series		
Height, h = _____		
n	Δx_n	v_n
0	—	—
1		
2		
3		
4		
5		
6		
7		
8		
9		
10		
11		
12		
13		
14		
15		
16		
17		
18		
19		
20		
21		

n	Δx_n	v_n
22		
23		
24		
25		
26		
27		
28		
29		
30		

n	Δx_n	v_n
22		
23		
24		
25		
26		
27		
28		
29		
30		

ANALYSIS For both series of measurements, calculate $v_n = \Delta x_n / \Delta t$, plot it against $t_n = n\Delta t$, and draw straight lines through both sets of points. Find the acceleration a in both cases from the slopes of the lines, and determine g from Eq. 5. Compare the average value of g to the expected value of 980 cm/s².

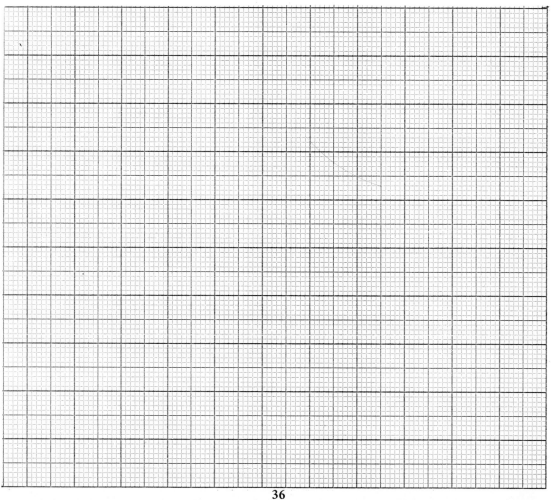

h	a	g

Average value of g = _____

QUESTIONS

1 What is the experimental uncertainty of an individual position measurement?

2 (*a*) Suppose the restraining string were cut at the same instant a spark jumped. How far would the glider move before the next spark?

  (*b*) From your first measurement, estimate the time interval between the instant the string was cut and the instant the glider reached position 0.

3 Using the data from one set of measurements, plot the distance of the glider from position 0 against time. Compare with Fig. 4.2

Part II   **Pulley and Plane**

DATA

Mass m = _____

Time interval Δt = _____

First Series

Mass, m' = _____

n	Δx_n	v_n
0	—	—
1		
2		
3		
4		
5		
6		
7		
8		
9		
10		
11		
12		

Second Series

Mass, m' = _____

n	Δx_n	v_n
0	—	—
1		
2		
3		
4		
5		
6		
7		
8		
9		
10		
11		
12		

n	Δx_n	v_n
13		
14		
15		
16		
17		
18		
19		
20		
21		
22		
23		
24		
25		
26		
27		
28		
29		
30		

n	Δx_n	v_n
13		
14		
15		
16		
17		
18		
19		
20		
21		
22		
23		
24		
25		
26		
27		
28		
29		
30		

ANALYSIS Repeat the analysis described in Part I.

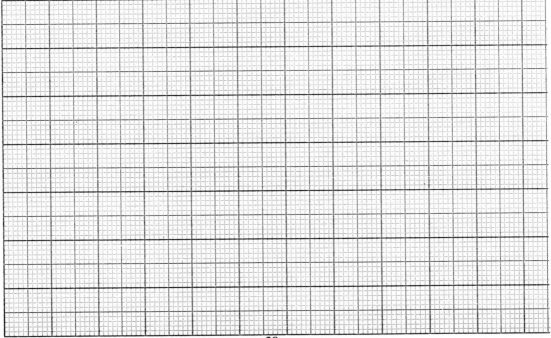

m'	a	g

Average value of g = _____

QUESTIONS

1. Calculate the *second differences* $\Delta v_n = v_{n+1} - v_n$ from the data from one set of measurements. The acceleration is defined as

$$a_n = \frac{\Delta v_n}{\Delta t}$$

Since a_n and Δt are constants, the values of Δv_n should all be the same within experimental uncertainty. Calculate a from the average value of Δv_n.

2. Suppose the mass of the glider is 300 g. What mass m' must be suspended from the pulley to give the system an acceleration of (a) $\tfrac{1}{2} g$, (b) $0.9\, g$.

EXPERIMENT 5 WORK AND ENERGY

GOALS

1 To demonstrate the work-energy theorem.

2 To measure the acceleration of gravity.

3 To become familar with electronic timing instrumentation.

EQUIPMENT

Linear air track   Electronic timer
Two photogates   Pulley and weights

INTRODUCTION

The work-energy theorem states that the total work W done on an object moving from point A to point B is equal to the difference $\Delta K = K_B - K_A$ in the kinetic energies of the object at A and B. The *work* done on the object is defined as

$$W = F_x d$$

where F_x is the magnitude of the component of the force parallel to the direction of motion of the object, and d is the distance the object is moved. The *kinetic energy* of an object is defined as

$$K = \tfrac{1}{2} m v^2$$

where m is the mass of the object and v is its speed. In terms of these equations, the work-energy theorem can be written

$$F_x d = \tfrac{1}{2} m v^2 - \tfrac{1}{2} m v^2 \qquad\qquad 1$$

In this experiment you will study this relationship for the same two air-track arrangements used in Experiment 4.

INCLINED PLANE Figure 5.1 shows an object of mass m sitting on an inclined plane. If the plane is frictionless, the magnitude of the total force on m parallel to the plane is

$$F_x = mg \sin\theta$$

so the work done on m in moving it from A to B, a distance d along the incline, is

$$W = F_x d = mgd \sin\theta = mgh$$

where h is the vertical distance between points A and B. From Eq. 1 we have

or
$$mgh = \tfrac{1}{2} m v_B^2 - \tfrac{1}{2} m v_A^2$$
$$v_B^2 - v_A^2 = 2gh \qquad\qquad 2$$

This equation says that the difference in the square of the speeds of an object at two fixed points on

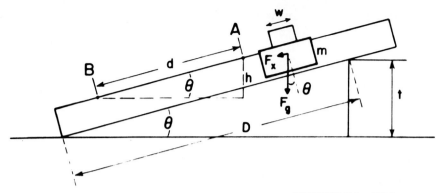

FIGURE 5.1 Glider on an inclined air track.

a frictionless inclined plane is the same regardless of the object's initial speed.

PULLEY AND PLANE Figure 5.2 shows an object of mass m sitting on a frictionless horizontal plane. The object is connected by means of a cord and pulley arrangement to a vertically suspended object of mass m'. The tension T of the cord is the only force on m parallel to the plane, so when m moves a distance d from point A to point B, the work done on it is Td. The total vertical force on m' is $m'g - T$, so when m' moves down a distance d, the work done on it is $(m'g - T)d$. Thus, the total work done on the two masses together is

$$W = Td + (m'g - T)d = m'gd$$

and it is equal to the change in the total kinetic energy of the two masses:

FIGURE 5.2 Glider on a frictionless horizontal plane connected by means of a cord and pulley to a vertically hanging mass.

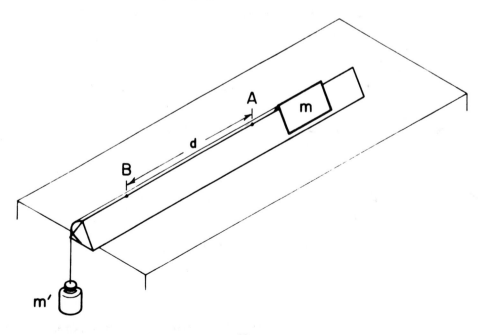

$$m'gd = K_B - K_A = (\tfrac{1}{2}mv_B^2 + \tfrac{1}{2}m'v_B^2) - (\tfrac{1}{2}mv_A^2 + \tfrac{1}{2}m'v_A^2)$$
$$= \tfrac{1}{2}(m+m')v_B^2 - \tfrac{1}{2}(m+m')v_A^2$$

or
$$v_B^2 - v_A^2 = \frac{2m'gd}{m+m'} \qquad\qquad 3$$

(Note that in deriving this equation we have used the fact that m and m' are connected, so that they travel the same distance and at the same speed.) As before, the difference in the square of the speeds at two fixed points is the same regardless of the initial speed, although in this case this difference depends on the masses of the objects, whereas it is independent of mass in the case of the inclined plane.

REFERENCE Cromer: *Physics for the Life Sciences*, Sec. 5.1, 5.2, and 5.3; Experiment 4 of this Manual.

Part I Inclined Plane

PROCEDURE

The apparatus consists of the linear air track used in Experiment 4, and two photogates to measure the speed of a glider at two fixed points on the track. A photogate is a photocell and light source connected to an electronic timer in such a way that whenever the light hitting the photocell is interrupted, the timer is turned on. The timer is turned off as soon as the light again hits the photocell. Two such photogates are placed across the air track at two points A and B. The light hitting the photocell is interrupted by a small flag of width w mounted on the glider. As the glider passes through each gate, the transit times t_A and t_B for the flag to pass points A and B is measured. From these measurements the speeds $v_A = w/t_A$ and $v_B = w/t_B$ of the glider at points A and B are calculated.

REMARK If only one electronic timer is available, it can be used to time two photogates. Your instructor will show you how to connect the two photocells in series to the timer so that when the light hitting either photocell is interrupted, the timer is turned on. To make a measurement with this arrangement, one partner launches the glider while the second partner watches the timer. As the glider passes through gate A, the second partner must quickly note the timer reading, t_A, and write it down. As the glider passes through gate B, the transit time t_B is added to t_A, so that the final reading on the timer is $t_A + t_B$. Thus, t_B is obtained by subtracting t_A from the final timer reading. The first partner must catch the glider after it passes through gate B to prevent it from bouncing back through the gate and restarting the timer. Practice this procedure a few times before taking data.

Set up a photogate at two points A and B near each end of the track. Level the track and test the arrangement of the photogates by launching the glider on the leveled track. The transit times t_A and t_B should be equal to within a millisecond or so. When everything checks out, elevate one end of the track several centimeters by placing a metal block under it.

Measure the width w of the flag and the distance d between the photogates using the following procedure. Carefully move the glider toward gate A, until the timer turns on, and read the position x_A of the front edge of the glider on the scale along the side of the track. It is important that x_A correspond to the exact position at which the timer turns on. Move the glider through the gate, and find the position x'_A of the front end of the glider at which the timer turns off. Do the same for gate B. Then the width w and the distance d are given by

$$w = x'_A - x_A = x'_B - x_B$$

and
$$d = x_B - x_A = x'_B - x'_A$$

The photowidth w obtained this way may differ by a few millimeters from the geometrical width of the flag. It is important that the photowidth of the gate be measured to within 1 mm.

A series of measurements consists of at least four measurements of the transit times of the glider through gates A and B. For each measurement, try to launch the glider with a different initial speed, and be sure to include measurements taken by launching the glider both up and down the track. Finally, measure the thickness t of the block placed under the track.

Take a second series of measurements with a block of a different thickness under the track.

Measure the distance D between the legs of the track.

ANALYSIS

Calculate v_A, v_B, and $v_B^2 - v_A^2$ for each measurement. The quantity $v_B^2 - v_A^2$ should be the same within the experimental uncertainty of the experiment for all measurements taken with the same value of t.

Calculate the difference h in the vertical heights of the track at A and B from the equation

$$h = d \sin \theta = d \frac{t}{D} \qquad 4$$

For both series of measurements, use Eqs. 2 and 4 to determine g from the average values of $v_B^2 - v_A^2$. Compare the average value of g with the expected value.

Part II Pulley and Plane

PROCEDURE

Level the air track and connect the glider by means of a cord to a mass m' hanging from the pulley at one end of the track. Set up a photogate at two points A and B along the track. Point B should be near the pulley, and point A should be at least 20 cm from the other end of the track. Measure the width w of the flag and the distance d between the photogates by the procedure described in Part I.

A series of measurements consists of at least four measurements of the transit times of the glider through gates A and B. For each measurement release the glider from a different point between A and the end of the track in order to vary the speed with which it passes through gate A. (Do not give the glider an initial speed, since this will cause the cord to sag, and the glider and m' will then not move at the same speed.) Measure the mass m'.

Take a second series of measurements using a different mass m'.

Measure the mass m of the glider.

ANALYSIS

Repeat the analysis of Part I, using Eq. 3 to calculate the acceleration of gravity.

Name _____

Date _____

REPORT SHEET
EXPERIMENT 5   WORK AND ENERGY

Part I   Inclined Plane

DATA

Width of flag, w = _____

Distance between photogates, d = _____

Distance between legs of track, D = _____

First Series
Thickness of block, t = _____

t_A	t_B	v_A	v_B	$v_A^2 - v_B^2$

Average value of $v_A^2 - v_B^2$

Second Series
Thickness of block, t = _____

t_A	t_B	v_A	v_B	$v_A^2 - v_B^2$

Average value of $v_A^2 - v_B^2$

ANALYSIS   Calculate v_A, v_B, and $v_A^2 - v_B^2$ for each measurement and enter on the data table. For each series of measurements, calculate the difference h in the heights of points A and B from Eq. 4, and use Eq. 2 to calculate g from the average value of $v_A^2 - v_B^2$. Compare the average value of g with the expected value.

h	g

Average value of g = _____

QUESTIONS

1. Measure the mass m of the glider. Calculate the potential energy U_A of the glider at A, assuming that $U_B = 0$. From one measurement of v_A and v_B, calculate the kinetic energies of the glider at points A and B. Compare the total energies of the glider at A and B.

2. Calculate g from the data of your first set of measurements, assuming a value of w that is 1 mm less than your measured value. By what percent does the value of g change for a 1 mm change in w? Is an error in the measured value of w a random or a systematic error?

Part II   Pulley and Plane

DATA

Width of flag, $w =$ _____

Mass of the glider, $m =$ _____

Distance between photogates, $d =$ _____

First Series				
Mass, $m' =$ _____				
t_A	t_B	v_A	v_B	$v_A^2 - v_B^2$

Average value of $v_A^2 - v_B^2 =$ _____

Second Series				
Mass, $m' =$ _____				
t_A	t_B	v_A	v_B	$v_A^2 - v_B^2$

Average value of $v_A^2 - v_B^2 =$ _____

ANALYSIS

Calculate v_A, v_B, and $v_A^2 - v_B^2$ for each measurement and enter in data table. For each series of measurements, use Eq. 3 to calculate g from the average value of $v_A^2 - v_B^2$. Compare the average value of g with the expected value.

m'	g

Average value of $g =$ _____

QUESTIONS

1 Derive Eq. 3 by equating the change in the potential energy of the system to the change in kinetic energy.

2 Why is a 2% error in the measurement of w more serious than a 2% error in the measurement of m'?

EXPERIMENT 6   MECHANICAL EQUIVALENT OF HEAT

GOALS

1. To demonstrate that mechanical work can increase the temperature of a system.
2. To measure the mechanical equivalent of heat.
3. To study Newton's law of cooling.

EQUIPMENT

Mechanical-equivalent-of-
   heat apparatus (Cavendish
   form)
Thermometer
Balance
Timer
Conical paper sheets

INTRODUCTION

The law of conservation of energy states that energy can neither be created nor destroyed, but only changed from one form to another. For example, as a block slides along a table some of its kinetic energy is transformed into heat energy through the action of friction. This change in the form of the energy is shown by an increase in the temperature of the block.

Let us consider the block and table to be a single system, and suppose further that an external agent applies a constant force $\mathbf{F}_a$ on the block, which is just sufficient to keep it moving at constant speed. The total force on this system is zero because there is an equal but opposite force on the table due to the floor (another agent external to the system). The force $\mathbf{F}_a$ does an amount of work on the system equal to $F_a d$, where d is the distance through which the block moves. Therefore, what is observed is that the temperature of the system increases when an external agent does work on it this way. This work is of course measured in terms of mechanical units which in the mks system is the joule, J.

$$1\ J = 1 N \times 1\ N-m$$

The temperature of a system can also be increased by heating the system. The heat added to any system is typically expressed in terms of a special unit, the *calorie*. One calorie is defined as the amount of heat required to raise the temperature of 1 g of water by 1 Celsius degree (°C) near 15°C. In general, the heat Q gained or lost by a mass m of any substance when the temperature changes by ΔT is

$$Q = cm\Delta t$$

where c is a quantity called the *specific heat*, which is characteristic of the substance. The cgs and mks units of c are cal/g-°C and kcal/kg-°C, respectively. (By definition the specific heat of water is 1 cal/g-°C.) For example, the heat required to raise the temperature of a 100-g copper block by 10°C is

$$Q = cm\Delta T = 0.092\ \text{cal/g-°C} \times 100\ \text{g} \times 10°C = 92\ \text{cal}$$

where $c = 0.092$ cal/g-°C is the specific heat of copper.

In the absence of any external agent doing work on a system, the increase in the internal energy ΔE of the system comes only from the addition of a quantity of heat Q. Thus, for a given temperature change ΔT, ΔE (in calories) is given by

$$\Delta E = Q = cm\Delta T \qquad 2$$

In general, however, the internal energy can also be affected by the performance of work on or by the system. This leads to the following statement of the law of conservation of energy†

$$\Delta E = Q + \frac{W}{J} = (cm\Delta T) + \frac{W}{J} \qquad 3$$

Here Q is the heat added to the system in calories (it is negative if the system loses heat), and W is the work done on the system (it is negative if W is done by the system†). The factor J, which expresses the constant relationship between mechanical and thermal units of energy (joules and calories), is called the *mechanical equivalent of heat*. In this experiment the value of J is determined by measuring the rise in temperature of a thermally isolated system, when a measured amount of work is done on it but no heat is added to it. From the temperature rise and the specific heat of the system, we can calculate the heat Q that would have given the same temperature change if no work were done. The mechanical equivalent of heat then is given by $J = Q/W$. Because the system is not perfectly isolated, there is some unavoidable loss of heat from the system to the surroundings due to radiation and conduction through the air. A correction will be made for this loss in the second part of the analysis.

REFERENCE Cromer: *Physics for the Life Sciences*, Sec. 5.5 and 11.2.

PROCEDURE

The mechanical-equivalent-of-heat apparatus (Fig. 6.1) consists of a conical brass cup that can be rotated by means of a known applied torque. A hollow brass cone fits inside the cup and is filled with water. Both cones are thermally well insulated from the surroundings by a calorimeter housing. A thermometer used in conjunction with a stirring rod determines the system temperature. The work done on the system is obtained through the use of a spring balance and rotation counter.

To begin, remove the inner cone and fill it about three-fourths full with cold tap water. Check to see that the water temperature is at least 4°C below room temperature. Weigh the cone plus water on the balance and record the data. Before inserting the inner cone back into the apparatus, put a piece of thin paper into place around the inner cone, as directed by your instructor.

After assembling the cones, place them inside the housing taking care to line up the pins with the mating holes. Adjust the cord attached to the aluminum disc so that it follows the circumferential groove about three-quarters of the way around before running under the pulley and up to the spring balance (Fig. 6.2). Insert the thermometer and stirring rod through the holes in the rubber stopper, and plug the stopper into its seat. Be sure that the thermometer bulb is fully immersed in the water.

During the experiment itself, the duties will be split along defined lines: One partner will crank the handle while at the same time watching the spring-gauge dial; it is *important* to try to maintain a constant value of the force at all time. The other partner will read the thermometer every thirty seconds, stirring the water before each reading. This partner will also serve as the record keeper.

---

†Equation 3 is also known as the *first law of thermodynamics*. For simplicity, the sign convention used here for W is the opposite of the convention traditionally used in thermodynamics, so that Eq. 3 differs by a minus sign from Eq. 11.8 in Chap. 11 of Cromer: *Physics for the Life Sciences*.

FIGURE 6.1 Mechanical-equivalent-of-heat apparatus [Cavendish form]. [*Sargent-Welch Scientific Co.*]

Before starting the experiment, it is a good idea for the cranker to practice. This procedure may heat the water enough so that fresh water must be obtained and weighed before the data run is started. As stated before, it is important that the initial water temperature be at least 4° below room temperature.

At the start of the run the following data should be recorded: room temperature T_R, the initial water temperature T_1, and the reading of the revolution counter. After cranking has started, the data taker will measure and record the water temperature every thirty seconds.

Cranking should be continued until the water temperature is about 8°C above room temperature. After the cranking is stopped, the water temperature must still be recorded every thirty seconds until it reaches a maximum value. Record both the temperature and the time when this occurs. Also record the constant value of the gauge-dial reading in newtons, and the final counter reading.

After the maximum temperature is reached, *continue* to measure the water temperature *every two minutes* for about fifteen minutes. Don't forget to stir before taking each reading. This data will be used in the second part of the analysis to determine the rate at which heat is lost to the surroundings.

ANALYSIS 1

Assuming that no loss of heat to the environment has occurred, a rough value of the mechanical equivalent of heat can be easily obtained. The work W done by the applied force F on the system is Fd, where d is the total distance through which the point of application moves. Because the system is in rotational equilibrium, the torque τ produced by F must be counterbalanced by the torque produced by the tension F' in the cord that runs around the periphery of the stationary disc. This torque is given, in fact, by the product of the reading on the spring balance and the radius of the disc. Thus, since

$$W = Fd = F 2\pi r = 2\pi \tau$$

per revolution (where r is the lever arm of the applied force), the work done can be computed by multiplying the torque by $2\pi n$, where n is the total number of revolutions.

More explicitly, the work done is given by

$$W = 2\pi n \tau = 2\pi n F' R \qquad 4$$

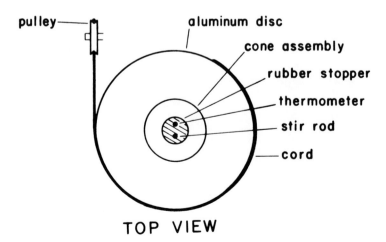

FIGURE 6.2 Detail of the mechanical-equivalent-of-heat apparatus, showing the attachment of cord to the aluminum disc on top of the apparatus.

where R is the radius of the disc to which the string is attached, and F' is the force measured on the spring balance.

The change in energy of the system produced by the work W is

$$\Delta E = \frac{W}{J} \text{ (in calories)} \qquad 5$$

if no heat enters or leaves the system. Conversely, the amount of heat needed to produce this energy change, if no work were done, is

$$Q = \Delta E = (m_w c_w + m_b c_b + m_s c_s + m_t c_t)\Delta T \text{ (in calories)} \qquad 6$$

where the m's are the masses and the c's are the specific heats of the water, the brass cones, the brass stirrer, and the thermometer, respectively. The temperature rise $\Delta T = T_2 - T_1$ is the difference between the *highest* temperature reached and the starting temperature of the system. The approximate value for the mechanical equivalent of heat J, is then given by

$$Q = \frac{W}{J} \text{ or } J = \frac{W}{Q} \qquad 7$$

Calculate J from your data and compare your value to the accepted value of 4.186 joules/cal. (The specific heats and weights of the components of the apparatus will be supplied by your instructor.)

ANALYSIS 2

We now turn to the problem of correcting the measurements for the heat lost to the surroundings. On a piece of graph paper make a plot of the water temperature T_w against elapsed time for the entire period during which observations were made. Draw a smooth curve through the data. This curve can be split up into two distinct parts: *1* the part where the system temperature is increasing, which is called the warming curve; *2* the part where the temperature is decreasing, which is called the cooling curve (Fig. 6.3).

During the time when the water temperature is less than room temperature heat flows from the surroundings into the system. Conversely, during the time when the water temperature is greater than room temperature, some heat flows into the surroundings from the system. It is also apparent that the lar-

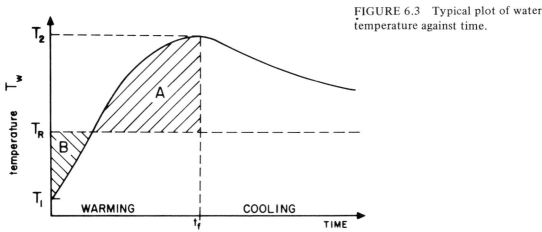

FIGURE 6.3 Typical plot of water temperature against time.

ger the difference between these two temperatures, the greater will be the rate of heat flow.

After the system has reached its maximum temperature it starts to cool by losing heat to the surroundings. *Newton's law of cooling* states that the rate of temperature change of a cooling body is proportional to the temperature difference $\Delta T = T_W - T_R$ between the system and its surroundings. As a consequence of this, ΔT decreases by the same *fraction* in equal time intervals. For example, if ΔT decreases to $\frac{1}{2}\Delta T$ in the time $t_{1/2}$, it will decrease to $\frac{1}{2}(\frac{1}{2}\Delta T) = \frac{1}{4}\Delta T$ in the time $t_{1/2} + t_{1/2} = 2 t_{1/2}$.

From the cooling curve, determine the time $t_{1/2}$ required for $\Delta T = T_W - T_R$ to decrease to one-half of its maximum value.

REMARK It may be helpful to construct a horizontal line on the plot which passes through the value of T_R in order to determine the difference $\Delta T = T_W - T_R$ more easily, as shown in Fig. 6.3.

If we assume that the heat flow follows the same law during the warming period, we can calculate the net amount of heat lost to the surroundings. One can obtain this correction by applying the following procedure. First, on the warming curve plot, construct a horizontal line which passes through the room temperature value T_R, as shown in Fig. 6.3. Compute the area of region A and region B by approximating them as triangles. Subtract area B from area A and divide this number by the elapsed time taken for the system to reach the maximum temperature, t_f. This quantity is the average temperature difference ΔT_{avg} between the system and the surroundings during the warming period. The net heat loss $-Q'$ during the warming period is related to the average temperature difference by the following equation

$$-Q' = \frac{t_f}{1.44 t_{1/2}} (m_w c_w + m_b c_b + m_s c_s + m_t c_t)\Delta T_{avg} \qquad 8$$

where as before, the m's are the masses and the c's are the specific heats of the various elements in the system, and t_f is the elapsed time taken for the system to reach its maximum temperature.

Because the system is not perfectly isolated during the warming period, it loses the heat $-Q'$ while the work W is being performed on it. Consequently, Eq. 5 must be corrected to read

$$\Delta E = \frac{W}{J} + Q' \qquad 9$$

Equating this to the heat required to produce the same temperature change, we obtain

$$\frac{W}{J} + Q' = Q$$

or

$$J = \frac{W}{Q - Q'} \qquad 10$$

Perform the calculation and compare the corrected value of J with the standard value quoted earlier.

Name _____

Date _____

REPORT SHEET
EXPERIMENT 6   THE MECHANICAL EQUIVALENT OF HEAT

Record the readings as indicated:

   Room temperature, T_R = _____ °C

   System temperature at start, T_1 = _____ °C

   Revolution-counter reading at start = _____

DATA

System temperature (°C)	Elapsed time (°C)

Revolution-counter reading at finish = _____

Maximum system temperature, T_2 = _____ °C

Elapsed time to reach T_2, t_f = _____ s.

Spring-gauge force, F' = _____ N

Radius of aluminum disc, R = _____ m

ANALYSIS 1

1 Calculate the work done on the system W:

   Number of revolutions, n = _____

   Work done, $W = 2\pi n F' R$ = _____ J

2 Calculate the thermal energy gained by the system ΔE:

   Total temperature change, $T_2 - T_1$ = _____ °C

Fill in the chart by making appropriate measurements and consulting references if necessary.

	Mass (g)	Specific heat (cal/g - °C)	Product
Water	$m_w =$	$c_w =$	$m_w c_w =$
Brass cones	$m_b =$	$c_b =$	$m_b c_b =$
Brass stirrer	$m_s =$	$c_s =$	$m_s c_s =$
Thermometer	$m_t =$	$c_t =$	$m_t c_t =$
		Sum =	

The system thermal energy is calculated from:

$$\Delta E = (m_w c_w + m_b c_b + m_s c_s + m_t c_t)(T_2 - T_1) = \underline{\hspace{3cm}} \text{ cal}$$

Determine the mechanical equivalent of heat (approximate)

$$J = \frac{W}{\Delta E} = \underline{\hspace{5cm}} \text{ J/cal}$$

QUESTIONS

1. Assuming that heat is actually lost to the surroundings, should your preceding value of J be higher or lower than the accepted value?

2. Which element of the system is most important in terms of the amount of thermal energy it absorbs for a given temperature difference?

ANALYSIS 2

1. On the graph, plot T_w against elapsed time, and draw a smooth curve through the data.

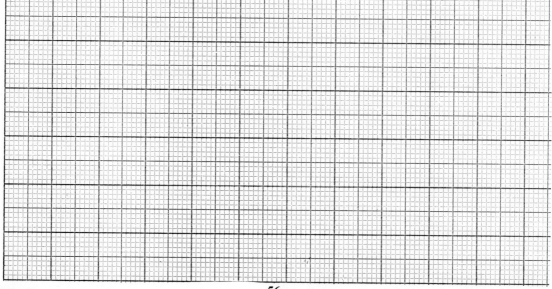

2 From the warming-curve plot, determine the following:

Elapsed warm-up time, $t_f =$ _____ s
Area $A =$ _____ °C - s
Area $B =$ _____ °C - s
Difference $A - B =$ _____ °C - s
Average temperature difference,

$$\Delta T_{avg} = \frac{(A - B)}{t_f} = \underline{\hspace{2cm}} \text{°C}$$

From the cooling-curve plot, determine the following:

Half time, $t_{1/2} =$ _____ s

3 Determine the net heat loss:

$$-Q' = \frac{t_f}{1.44 \, t_{1/2}} (m_w c_w + m_b c_b + m_s c_s + m_t c_t) \Delta T_{avg} = \underline{\hspace{2cm}} \text{cal}$$

4 Calculate the corrected value for J:

$$J' = \frac{W}{(\Delta E - Q')} = \underline{\hspace{2cm}} \text{J/cal}$$

QUESTIONS

1 Explain why it is best to start this experiment at a temperature somewhat below room temperature.
2 Why is it necessary for the data taker to stir the water before each reading?
3 Suppose cold water were placed in the apparatus and left there with no work being done. Describe how the temperature of the system would change as time passed.
4 If a person sitting in a room at a temperature of 20°C losses heat at the rate of 110 kcal/h, what would be the rate of heat loss in a room at a temperature of 18°C? (The temperature of the body is 37°C.)

properties
of
matter

EXPERIMENT 7   PRESSURE IN FLUIDS

GOALS

1 To demonstrate Pascal's law using a hydraulic lift.

2 To learn to use a manometer to measure pressure.

3 To measure the density of mercury.

EQUIPMENT

Small mercury-filled manometer
Plastic syringes with platforms mounted on the plungers
Tubing
Weights

For best results, the tubing should be securely connected to the syringes with plastic connectors and the syringes should be firmly clamped to stands.

INTRODUCTION

Pressure is the force per unit area exerted on or by a surface. Thus, the pressure p that a force F exerts on an area A is

$$p = \frac{F}{A}$$

In the mks system, the unit of force is the newton, N, and the unit of area is the square meter, m^2, so that the unit of pressure is the newton per square meter, N/m^2.

The concept of pressure is especially useful in the study of fluids (liquids and gases) because of *Pascal's law*, which states that, neglecting the effect of gravity, *the pressure in a fluid at rest is the same everywhere in the fluid*. This is important because it means that in a given situation, a fluid is characterized by a single value of p. Pascal's law is studied in Part I of this experiment.

When gravity is taken into account, it is found that the pressure below the surface of a fluid increases with depth. That is, the pressure p_1 at one point in a fluid is greater than the pressure p_2 at a point above it (Fig. 7.1). This is due to the weight of the additional fluid between the two points. The pressures p_1 and p_2 are related by the equation

$$p_1 - p_2 = \rho g h \qquad \qquad 1$$

where ρ is the density of the fluid, g is the acceleration of gravity (9.8 m/s^2), and $h = h_2 - h_1$ is the difference in the vertical heights of the two points. (The *density* of a substance is its mass per unit volume, and its mks unit is kilograms per cubic meter, kg/m^3.) The pressure in a fluid is the same, however, at all points that are the same depth below the surface, regardless of the shape of the fluid.

Pressure is measured with a *manometer*, a U-shaped tube half filled with a liquid (usually water or mercury). One end of the tube is open to the atmosphere, and the other end is connected to the ves-

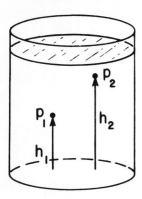

FIGURE 7.1 Pressure at two points in a fluid.

sel whose pressure is to be measured, as shown in Fig. 7.2. The pressure at point A is equal to the pressure p in the vessel, because both are at the same elevation. The pressure p_1 at point B is related to the pressure p_2 at the top of the open column by Eq. 1. But p is equal to p_1, because points A and B are at the same elevation, and p_2 is equal to the pressure of the atmosphere p_0, so that we have

$$p = p_0 + \rho g h \qquad\qquad 2$$

where $h = h_2 - h_1$ is the difference in the heights of the liquid in the two columns.

REFERENCE Cromer: *Physics for the Life Sciences,* Sec. 7.2, 7.3, and 6.2.

Part I Pascal's Law

PROCEDURE

The apparatus consists of two water-filled plastic syringes of different diameters connected by a tube (Fig. 7.3). Press down on the plunger of one syringe while your partner presses down on the other. Note that the one who presses on the smaller syringe can always lift the larger syringe. This illustrates the operation of a hydraulic lift, which is a simple machine based on Pascal's law.

With no mass placed on the large syringe, add masses to the small syringe until its plunger just

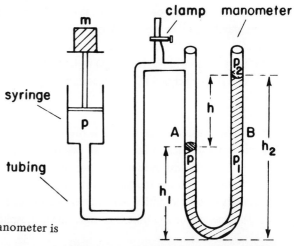

FIGURE 7.2 A manometer. One arm of the manometer is connected to a syringe.

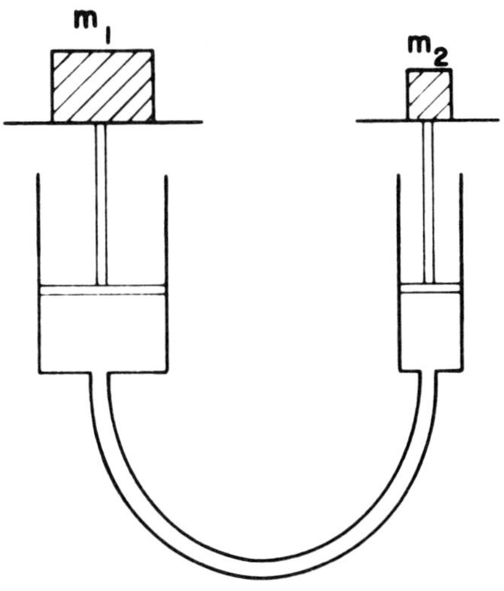

FIGURE 7.3 Two syringes of different sizes connected by a tube.

starts to fall. This condition occurs when the plunger falls steadily if given a slight push, but remains at rest otherwise. Determine to the nearest 50 g the mass m_2 needed to make the small plunger fall. Repeat the measurement with 500-g, 1-kg, 2-kg, and 3-kg masses on the large syringe.

If the diameters of the syringes are not given, measure the diameters of the duplicate syringes with a caliper.

ANALYSIS

Plot m_1, the mass on the large syringe, against m_2, the mass on the small syringe, and draw a straight line through the points. (Because of the friction in the syringes, this line will not pass through the origin.)

Mass m_1 exerts a force $m_1 g$ on the large syringe. Therefore, if we neglect the friction between the plunger and the walls of the syringe, the pressure exerted on the fluid by the plunger of the large syringe is

$$p_1 = p_0 + \frac{m_1 g}{A_1}$$

where A_1 is the cross sectional area of the large syringe, and p_0 is the pressure due to the atmosphere. Likewise, the pressure exerted on the fluid by the plunger of the small syringe is

$$p_2 = p_0 + \frac{m_2 g}{A_2}$$

According to Pascal's law, these pressures are equal, since they are exerted on the same fluid. Consequently we have

$$p_0 + \frac{m_1 g}{A_1} = p_0 + \frac{m_2 g}{A_2}$$

or

$$\frac{m_1}{m_2} = \frac{A_1}{A_2} \qquad 3$$

which says that the ratio of the masses, and hence the ratio of the forces applied to the two pistons is equal to the ratio of the areas. This is the principle of the hydraulic lift: a small force applied to a small area can lift a large force applied to a large area.

The *mechanical advantage* M of a hydraulic lift is the ratio of the large force to the small force

$$M = \frac{m_1 g}{m_2 g} = \frac{m_1}{m_2}$$

Therefore, we have

$$m_1 = M m_2$$

so that M is equal to the slope of the straight line drawn through the data points. Find the slope from your graph.

From Eq. 3 we see that in the ideal (frictionless) case, the mechanical advantage is equal to A_1/A_2. Calculate the ideal mechanical advantage from the diameters of the cylinders, and compare it with the slope of the line.

The *efficiency* e of a simple machine is the ratio of the actual mechanical advantage to the ideal mechanical advantage $M = A_1/A_2$.

$$e = \frac{m_1/m_2}{M}$$

Calculate e for each measured value of m_1, using the experimental value of m_1 and m_2, and the ideal mechanical advantage M.

Part II The Manometer

PROCEDURE

The apparatus consists of a plastic syringe connected by a tube to one arm of a mercury-filled manometer (Fig. 7.2). The other arm of the manometer is open to the atmosphere. The manometer will be used to measure the pressure in the syringe when masses of different magnitudes are placed on the plunger.

WARNING: Mercury is a cumulative poison, which can be absorbed through the skin. Should the manometer break, do not play with the mercury. Notify your instructor so that he can properly dispose of the mercury.

Open the clamp on the tube to equalize the pressure on both sides of the manometer. Lift the plunger halfway up the syringe, and hold it there while you close the clamp. Place a 500-g mass on the plunger and record the mercury levels in both columns.

Add masses onto the plunger in 100-g increments, up to a maximum of 1000 g. Record the mercury levels for each increment.

REMARK Because of friction in the syringe, it is necessary that each successive measurement be made with a heavier mass. Therefore, do not remove a mass to check an earlier measurement. If an error is made, the entire sequence must be repeated starting with the equalization of the pressure on both sides of the manometer.

If the diameter of the syringe is not given, measure the diameter of the duplicate syringe with a caliper.

ANALYSIS

The pressure p in the syringe due to the pressure p_0 of the atmosphere and the weight mg on the plunger is

$$p = p_0 + \frac{mg}{A}$$

where A is the cross sectional area of the syringe.

Equation 2 gives the pressure p in the syringe in terms of the density of mercury ρ and the height difference h of the mercury columns. Equating these two expressions for p we get

$$p_0 + \frac{mg}{A} = p_0 + \rho g h$$

or

$$h = \frac{m}{\rho A} \qquad 4$$

Plot h against the mass m on the plunger. (Be sure to express m in kilograms and h in meters.) We see from Eq. 4 that the slope of the straight line through the points is equal to $1/\rho A$. (Because of friction in the syringe, the line will not pass through the origin.) From the slope of the line and the diameter of the syringe, calculate the density of mercury. (Be sure to express A in square meters.)

Name_____

Date_____

REPORT SHEET
EXPERIMENT 7   PRESSURE IN FLUIDS

Part I   Pascal's Law

DATA

Mass on large syringe m_1	Mass on small syringe m_2

Diameter of large syringe, d_1 = _____

Diameter of small syringe, d_2 = _____

ANALYSIS

1 Plot m_1 against m_2.

2 The experimental value of the mechanical advantage M is equal to the slope s of the straight line drawn through the data.

$s = \underline{\hspace{2cm}}$

3 The ideal mechanical advantage is

$M = \dfrac{A_1}{A_2} = \underline{\hspace{2cm}}$

4 Tabulate the efficiency e for each value of m_1, using the equation

$e = \dfrac{m_1/m_2}{M}$

m_1	e

QUESTIONS

1 Note that the best straight line through your data does not pass through the origin. Why?

2 Why does the slope of the line give a much better value for M than is obtained by averaging the measured values of m_1/m_2?

3 Why does the efficiency increase as m_1 increases?

Part II   The Manometer

DATA

Mass	Height of mercury levels		Height difference
m	h_1	h_2	h

Diameter of the syringe, $d = \underline{\hspace{2cm}}$

ANALYSIS

1 Plot h against m.

2 Calculate the density of mercury from these data.

$\rho =$ _____

QUESTIONS

1 Suppose a syringe with half the diameter of the one used in this experiment were connected to a mercury manometer. What would h be when a 700 g mass was put on the plunger? (Neglect friction.)

2 Why must the masses be added in succession in this experiment?

3 Find the conversion from newtons per square meter (N/m²) to millimeter of mercury (mm Hg) from the value of the density of mercury found in this experiment. (The actual conversion is 1 mm Hg = 133 N/m².)

4 A syringe with a diameter of 1.4 cm is used to give a hypodermic injection directly into an artery. If the maximum arterial blood pressure is 120 mm Hg, what is the minimum force that must be applied to the plunger of the syringe to inject the contents of the syringe into the artery?

EXPERIMENT 8   FLUID FLOW

GOALS

1 To measure the flow of water through a capillary at different pressures.

2 To measure the resistances of capillaries of different diameters and lengths.

3 To measure the resistance of two capillaries connected in series.

EQUIPMENT

Two constant-level cups
Graduated cylinder
Glass capillaries (two 1-mm capillaries and one 2-mm capillary)
Timer
Tubing
Clamps
Stands
Beakers

INTRODUCTION

A fluid flows through a pipe because of the pressure difference between the ends of the pipe. The *fluid flow Q* is the volume of fluid that flows through the pipe in one second. Thus, if V is the volume of fluid that flows through a pipe in time t, the fluid flow Q is

$$Q = \frac{V}{t} \qquad 1$$

For a given pipe carrying a given fluid, Q increases as the pressure difference $p_1 - p_2 = \Delta p$ between the ends of the pipe increases.

This experiment studies the relation between Q and Δp for water flowing through narrow pipes (capillaries). When the flow rate is sufficiently small, Q is proportional to Δp, so that we can write

$$Q = \frac{\Delta p}{R}$$

or

$$\Delta p = RQ \qquad 2$$

where R is a constant, called the *resistance*. This relation is known as *Poiseuille's law*. It is found to be valid for water whenever Q (measured in ml/s) is less than the radius of the capillary (measured in mm). When Q becomes larger than this, the flow becomes turbulent, and the resistance increases.

The resistance depends on the *viscosity v* of the fluid, and on the length L and radius r of the capillary. For nonturbulent (*laminar*) flow, R is given by

$$R = \frac{8vL}{\pi r^4} \qquad 3$$

The important points to note about this formula are that R is proportional to L and is inversely proportional to the fourth power of r. This means that, if L is doubled, then R is doubled, and if r is doubled, then R is decreased by the factor

$$\frac{1}{2^4} = \frac{1}{16}$$

Consequently, relatively small changes in the radius of a capillary cause large changes in its resistance and, hence, large changes in the fluid flow through the capillary.

REFERENCE Cromer: *Physics for the Life Sciences*, Sec. 7.5.

Part I Resistance of a Capillary

PROCEDURE

The apparatus consists of a capillary connected by tubes to the drains of two constant-level cups (Fig. 8.1). When the cups are at different heights, a pressure difference Δp is produced between the ends of the capillary. This pressure difference is equal (in centimeters of water, cm H_2O) to the difference in height $\Delta h = h_1 - h_2$ of the water level in the cups. The water level in each cup is kept constant by means of an overflow pipe.

To begin, connect the shorter of the two 1-mm capillaries to tubes A and B (Fig. 8.1). Place a beaker under each cup, and fill both cups with water. Squeeze tubes A and B several times to remove all the air bubbles that are trapped in them. When the air bubbles are removed, water should drip into beaker II at a steady rate.

To take a measurement, clamp tube A, fill both cups to the top of the overflow, and empty beaker II. Remove the clamp but keep the tube pinched closed with your finger. Then start the timer and release the tube simultaneously. As water flows into beaker II, keep adding water to cup I so that its level remains at the top of the overflow. Allow the flow to continue for approximately five minutes, or until you have collected approximately 50 ml of water, whichever is sooner. To stop the flow, clamp tube A and stop the timer simultaneously.

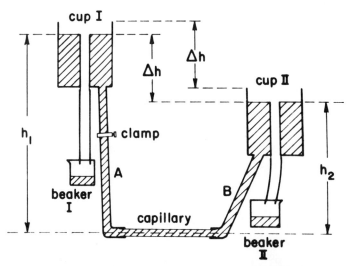

FIGURE 8.1 Schematic diagram of the fluid-flow apparatus.

Measure the volume V of water collected in beaker II by pouring it into a graduated flask. Record V together with the height difference Δh of the cups and the elapsed time t. Do this measurement for four different values of Δh, ranging from 5 cm to 40 cm.

REMARK If the capillary is kept in a horizontal position on the table, Δh is equal to the vertical distance between the top edges of the cups (Fig. 8.1). It may be easier to measure this distance directly, than to measure h_1 and h_2 separately.

Repeat the measurement with the longer 1-mm capillary. Take three measurements with Δh between 10 cm and 40 cm.

Repeat the measurement with the 2-mm capillary. Take six measurements with Δh between 3 cm and 20 cm. Be sure that three measurements are made with Δh less than 10 cm. The flow rate will be much faster with the 2-mm capillary than with the 1-mm capillary, so be sure to keep adding water to cup I to maintain the water level constant.

REMARK To avoid excessive spillage, please observe the following procedure for changing capillaries:

1 Clamp tube A close to the capillary.

2 Disconnect the capillary from tube A, and let tube A drain into a beaker.

3 Hold tube B above cup II and disconnect the other end of the capillary.

4 Let tube B drain into a beaker. You are now ready to connect the next capillary and refill the cups.

ANALYSIS

From Eq. 1 calculate the fluid flow Q for each measurement. For each capillary, plot Δh against Q, and draw a straight line from the origin through these points. Determine the resistance R of each capillary from the slope of the line.

The diameters of the 1-mm and 2-mm capillaries are approximately 1 mm and 2 mm, respectively. Measure the length L of each capillary and calculate the quantity $r^4 R/L$. According to Eq. 3 you should get the same value for each capillary. There will be considerable discrepancy however, because the diameters of the capillaries are not exactly 1 mm and 2 mm.

Part II Resistance of Two Capillaries in Series

PROCEDURE

Connect together the two 1-mm capillaries with a short piece of tubing. Measure the fluid flow Q through this combination for several values of Δh and record the data.

ANALYSIS

Plot Δh against Q, and determine the combined resistance R of the two capillaries from the slope of the straight line drawn through the data points. The total resistance R of two capillaries in series is

$$R = R_1 + R_2 \qquad\qquad 4$$

where R_1 and R_2 are the resistances of the individual capillaries. Calculate R from this equation and the values of the resistances found in Part I. Compare with the measured value. Electric resistance adds in the same way.

Name_____

Date_____

REPORT SHEET
EXPERIMENT 8   FLUID FLOW

Part I   Resistance of A Capillary

DATA

Capillary r L	Difference in height Δh (cm)	Volume V (ml)	Time t (s)	Fluid flow Q (ml/s)
1				
2				
3				

ANALYSIS

1. Calculate the fluid flow Q and enter on data table. Plot Δh against Q for each capillary. Use different marks (e.g., X, O, Δ) to plot the data for different capillaries.
2. Determine the resistance of each capillary from the slope of the straight line through the data.
3. Calculate the quantity $r^4 R/L$ for each capillary.

Capillary	R	$r^4 R/L$
1		
2		
3		

QUESTIONS

1. In what unitl is the resistance measured in this experiment?
2. The diameter of a 1-mm capillary is between 0.8 mm and 1.2 mm. Calculate the quantity $r^4 R/L$ for the short 1-mm capillary, assuming its diameter is 0.8 mm. Compare this to the value you found when you took the diameter to be 1.0 mm.
3. Do all the measured points for the 2-mm capillary lie on a straight line? Explain any systematic departures from linearity.
4. The pressure of the blood changes by 20 mm Hg when it flows through a particular capillary.
   (*a*) If the fluid flow through this capillary is 4×10^{-9} cm^3/s, what is the resistance of the capillary?
   (*b*) During exercise the diameter of the capillary increases by a factor of 1.6 and the pressure change increases by a factor of 1.3. What is the fluid flow through the capillary in this case?

Part II   Resistance of Two Capillaries in Series

DATA   Two 1-mm capillaries in series:

Difference in height Δh (cm)	Volume V (ml)	Time t (s)	Fluid flow Q (ml/s)

ANALYSIS

1. Calculate the fluid flow Q and enter on data table. Plot Δh against Q.

2 Determine the resistance from the slope of the straight line drawn through the data. Compare this experimental determination of R to the theoretical value given by Eq. 4.

$R_{exp} =$ _____

$R_{th} = R_1 + R_2 =$ _____

QUESTIONS

1 Estimate the resistance of the tubing used to connect the capillaries to the cups. Show that it is negligible compared to the resistance of the capillaries.

2 Blood flows through the body at the rate of 85 ml/s under a pressure difference of 100 mm Hg. Calculate the total effective resistance of the circulatory system of the body in the same units in which the resistances of the glass capillaries are measured.

EXPERIMENT 9 BOYLE'S LAW

GOALS

1 To learn to read a barometer.

2 To learn the difference between gauge and absolute pressure.

3 To demonstrate Boyle's law by varying the pressure of a fixed quantity of gas.

EQUIPMENT

Flexible-tube manometer
Barometer, wall-mounted

INTRODUCTION

The pressure p, volume v, and (absolute) temperature T of an ideal gas are related by

$$pV = nRT \qquad 1$$

where n is the number of moles of gas and R is the universal gas constant. For a given quantity of gas at constant temperature, the right-hand side of Eq. 1 is a constant K, so we have

$$pV = K \qquad 2$$

or

$$V = \frac{K}{p} \qquad 3$$

In words, this last equation says that *the volume of a given quantity of gas at constant temperature is inversely proportional to its pressure*. This is *Boyle's law*. Note that Boyle's law is just a special case of the general gas law (Eq. 1).

In this experiment, Boyle's law is tested by varying the pressure of a fixed quantity of gas enclosed in a glass tube. The pressure p in Eq. 1 is the absolute pressure of the gas. It is equal to the sum of the gauge pressure $\bar{p}$, which is measured with a mercury-filled manometer, and atmospheric pressure p_0, which is measured with a barometer:

$$p = \bar{p} + p_0 \qquad 4$$

REFERENCE Cromer: *Physics for the Life Sciences*, Sec. 8.1, 8.2, and 8.3.

PROCEDURE

The apparatus consists of a flexible-tube manometer, which is a manometer composed of two glass arms connected by a flexible tube. One arm of the manometer is sealed at the top by a stopcock (Fig. 9.1). The quantity of gas to be investigated is enclosed between the stopcock and the mercury column in the arm. *Under no circumstances must you touch the stopcock.* The gauge pressure (in millimeters of mercury) of the enclosed gas is equal to the differences in the heights h_1 and h_2 (in millimeters) of the mercury columns in the two arms:

$$\bar{p} = h_2 - h_1 \qquad 5$$

Note that $\bar{p}$ is negative when h_1 is greater than h_2.

The volume of gas is equal to the product of the cross sectional area A of the closed tube and the length l of the gas column

$$V = Al$$

The tube is carefully made to have the same cross-sectional area everywhere, so that the length

$$l = h_0 - h_1 \qquad 6$$

is proportional to the volume of the gas. Since we are interested in testing a proportionality (Eq. 3), we can use l as a measure of the volume of the gas.

The gauge pressure $\bar{p}$ is varied by raising and lowering the two arms of the manometer. Take readings of h_0, h_1, and h_2 for at least ten different values of $\bar{p}$. Be sure to include some negative values of $\bar{p}$.

During the period your instructor will show you how to read the wall-mounted barometer. Record the barometric pressure in the same unit that you record the gauge pressure (millimeters of mercury).

ANALYSIS

Complete the entries in the data table. Note that the product pV is constant within the accuracy of the experiment.

Plot V against $1/p$. According to Eq. 3, the data points should lie on a straight line.

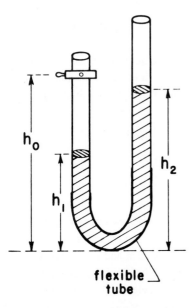

FIGURE 9.1   Flexible-tube manometer.

Name_____

Date_____

REPORT SHEET
EXPERIMENT 9   BOYLE'S LAW

DATA

Barometric pressure, p_0 = _____ mm Hg

Height of the mercury columns h_0 h_1 h_2	Volume V $(l = h_1 - h_2)$	Gauge pressure $\bar{p}$ $(h_2 - h_1)$	Absolute pressure p $(\bar{p} + p_0)$	pV	$\dfrac{1}{p}$

ANALYSIS

1 Complete the entries in the data table.

2 Plot V against $1/p$.

QUESTIONS

1 What gauge pressure is required to produce a volume that is half the smallest volume you obtained?

2 What gauge pressure is required to produce a volume that is twice the largest volume you obtained?

3 What would be the volume of the gas in this experiment at a gauge pressure of 120 mm Hg?

4 A patient is breathing oxygen from a cylinder which originally contained 16 l of oxygen at a pressure of 2700 lb/in.^{2}. (*a*) What is the volume of this gas at atmospheric pressure (14.7 lb/in.^{2})? (*b*) How long will this cylinder last a patient who is breathing 6 l of oxygen per minute? (*c*) What is the pressure of the gas in the cylinder after 2 hrs?

EXPERIMENT 10 TEMPERATURE

GOALS

1 To demonstrate that the pressure of a gas varies linearly with temperature.

2 To determine the temperature of absolute zero on the Celsius scale.

EQUIPMENT

Flexible-tube manometer
   with glass bulb attachment
Barometer, wall-mounted
Ice

Thermometer
Large beaker and stand
Bunsen burner

INTRODUCTION

The pressure p, volume V, and (absolute) temperature T of an ideal gas are related by

$$pV = nRT \qquad\qquad 1$$

where n is the number of moles of the gas and R is the universal gas constant. For a given quantity of gas in a fixed volume, the pressure is proportional to the absolute temperature

$$p = \frac{nRT}{V} = KT \qquad\qquad 2$$

where $K = nR/V$ is a constant. This last equation can in fact be taken as the definition of the absolute temperature T. In this experiment we shall see how this equation is used to relate the absolute temperature scale to the Celsius temperature scale.

The Celsius temperature scale is defined by arbitrarily assigning "temperature values" to a mixture of ice and water (the *ice point*) and to boiling water at 760 mm Hg (the *steam point*). The ice point is assigned the value of 0 temperature units, and the steam point is assigned the value of 100 temperature units. These units are called *degrees* and are usually abbreviated °C. Thus, the temperature of the ice point is 0°C, and the temperature of the steam point is 100°C.

A *thermometer* is used to measure other temperatures. A thermometer is an instrument that contains a substance with a physical property that changes with temperature. For example, a mercury thermometer contains mercury sealed in a glass tube with a bulb in one end. The mercury in the bulb expands when heated and contracts when cooled. The height of the mercury column in the tube is taken to be proportional to the temperature. For instance, when the mercury column is 1/5 of the distance between the 0°C and the 100°C marks, the temperature is 20°C.

This is satisfactory for most purposes, but in principle it ties the definition of temperature too closely to a special property of a particular substance. A more fundamental definition of temperature is obtained by using the pressure of an ideal gas, in accordance with Eq. 2.

In this experiment you will calibrate a gas thermometer, and compare it to a mercury thermometer.

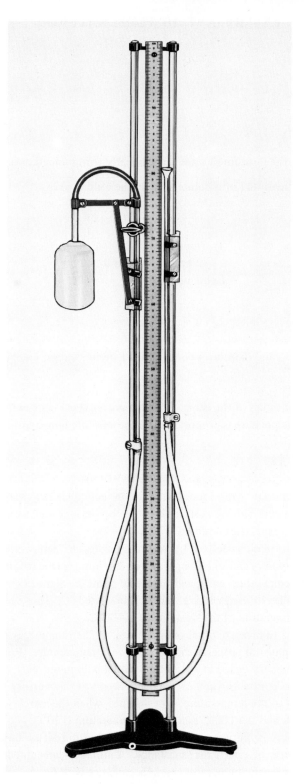

FIGURE 10.1 Flexible-tube manometer connected to a glass bulb. [*Sargent-Welch Scientific Co.*]

In the process, you will find the relation between the absolute and Celsius temperature scales, and determine the temperature of absolute zero on the Celsius scale.

REFERENCE Cromer: *Physics for the Life Sciences,* Sec. 8.2 and 8.3.

PROCEDURE

The apparatus consists of a glass bulb connnected to one arm of a flexible-tube manometer (Fig. 10.1). The stopcock connecting the bulb to the manometer is open. *Under no circumstances must you touch the stopcock.* Before each measurement the manometer arms are adjusted until the mercury level in the bulb-side arm is even with the bottom of the stopcock. In this way, the volume of gas is kept constant while the temperature and pressure of the gas in the bulb is varied.

To begin, adjust the heights of the two arms of the manometer until the mercury level is even with the bottom of the stopcock. Record the heights h_1 and h_2 of the mercury columns in the two arms, and the room temperature t as measured on a mercury thermometer.

Next, lower the bulb-free arm until the mercury in the bulb-side arm is about 10 cm below the stopcock. Then immerse the bulb in a beaker filled with ice and water. Attach the beaker to a stand, *taking care not to break the bulb*. Note that as soon as the bulb is immersed, the mercury level rises in the bulb-side arm. When the mercury stops moving, adjust the arms so that the mercury level is even with the bottom of the stopcock. Record the heights of the mercury columns. By definition, the temperature in this case is 0°C.

Lower both arms as far as they will go, taking care not to break the glass tubes. (Be sure to lower the bulb-free arm along with the bulb-side arm, so that the mercury does not rise above the level of the stopcock.) Immerse the bulb in a beaker of tap water, put a thermometer in the water, and place a Bunsen burner under the beaker. Before lighting the burner, have your instructor check that everything is in order.

As the water is heated, the mercury column in the bulb-side arm will fall. Raise the bulb-free arm until the mercury level in the bulb-side arm is even with the bottom of the stopcock. Record h_1, h_2, and t at several temperatures. Remember, a reading is valid only when the mercury level is even with the bottom of the stopcock. Also, be sure the thermometer is not touching the bottom of the beaker when you read it. The final reading is taken when the water is boiling vigorously. By definition, the temperature in this case is 100°C.

REMARK The temperature of boiling water is 100°C only when atmospheric pressure is 760 mm Hg. If the atmospheric pressure in your laboratory differs substantially from 760 mm Hg, use the temperature of boiling water given in Table 10.1.

TABLE 10.1

Atmospheric pressure (mm Hg)	Boiling point of water (°C)	Atmospheric pressure (mm Hg)	Boiling point of water (°C)
600	93.5	700	97.7
620	94.4	720	98.5
640	95.3	740	99.3
660	96.1	760	100.0
680	96.9	780	101.4

After you have completed your measurement at the boiling point, turn off the burner, lower the bulb-free arm, and remove the beaker. *Do not remove the beaker before you lower the bulb-free arm.* Measure the atmospheric pressure on the wall-mounted barometer.

ANALYSIS

Complete the entries in the data table. Prepare a graph, as shown in Fig. 10.2. The vertical axis goes from $-350°C$ to $+100°C$, and the horizontal axis goes from 0 to 1200 mm Hg. Plot the absolute pressures at the ice point (0°C) and steam point (100°C). Connect these two points by a straight line and extend the line until it intersects the vertical axis.

This straight line gives the relation between the temperature scale defined by a gas thermometer and the pressure of the gas in the bulb. Plot the rest of your data on the graph. The intermediate temperatures between 0°C and 100°C were measured with a mercury thermometer, so in principle they need not lie on the line. However, within the accuracy of these measurements, the mercury thermometer agrees with the gas thermometer, so your points should lie on the line.

The unit of temperature on the absolute (or *Kelvin*) temperature scale is the *kelvin* (K), and there are 100 K between the ice and steam points. The Kelvin temperature T is defined in terms of the pressure p in an ideal gas by

$$T = a \frac{p}{p_i} \qquad\qquad 3$$

where p_i is the pressure at the ice point. Determine a from the slope of the line and the measured value of p_i. (Careful measurements give $a = 273.15\ K$.) At the ice point $p = p_i$, so from Eq. 3 we see that at the ice point $T = a$. That is, a is the temperature of the ice point on the absolute scale. Since the temperature t of the ice point on the Celsius scale is 0°C, the two scales are related by

$$T = t + a = t + 273.15\ K \qquad\qquad 4$$

Absolute zero ($T = 0$) occurs when the gas pressure p is zero. Its value on the Celsius scale is given by the intersection of the straight line with the vertical axis. Determine the value of absolute zero from your graph. (The value of absolute zero from careful measurements is $-273.15°C$.)

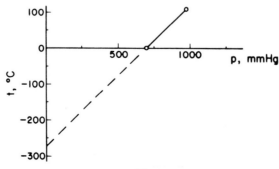

FIGURE 10.2 Plot of temperature against pressure, extrapolated to zero pressure.

Name_____

Date_____

REPORT SHEET
EXPERIMENT 10 TEMPERATURE

DATA

Barometric pressure, p_0 = _____ mm Hg

Heights of the mercury columns h_1 h_2	Gauge pressure p $(h_2 - h_1)$	Absolute pressure p $(\bar{p} + p_0)$	Temperature t

ANALYSIS

1. Complete the entries in the data table.
2. Plot p against t. Connect the ice and steam points by a straight line.

3 From the slope of the line determine the constant *a* in Eq. 3.

   a = _____

4 From the intersection of the line with the vertical axis, determine the value of absolute zero.

   t (absolute zero) = _____

QUESTIONS

1 Show that absolute zero is equal to $-a$, where a is the constant in Eq. 3.

2 What would be the gauge pressure of the gas in the bulb at a temperature of 150°C?

3 When the bulb is immersed in a beaker of hot water, the gauge pressure is 160 mm Hg. What is the temperature of the water?

EXPERIMENT 11   VAPOR PRESSURE

GOALS

1 To learn to apply Dalton's law of partial pressures.

2 To measure the vapor pressure of a liquid at different temperatures.

EQUIPMENT

Flexible-tube manometer
  with glass bulb attachment and T-connector (Fig. 11.1).
Clamps
Large beaker and stand

Thermometer
Ice
Tetrachloromethane
  (carbon tetrachloride)

INTRODUCTION

When a liquid is placed in a closed vessel the liquid evaporates until the partial pressure of its vapor in the vessel reaches a certain value p_v, called the *vapor pressure*. At this pressure, the vapor recondenses into liquid at the same rate as the liquid evaporates, so that a dynamic equilibrium is established between the liquid and its vapor. The vapor pressure at a given temperature is a characteristic property of the liquid, but the vapor pressure of all liquids increases rapidly with temperature. A liquid boils when its vapor pressure equals atmospheric pressure.

Consider a sealed vessel containing air at a pressure p_{air}. If a small quantity of liquid is introduced into the vessel, without any air entering or leaving, the partial pressure of the air will remain p_{air}. However, the liquid will evaporate until the partial pressure of its vapor is equal to the vapor pressure p_v of the liquid. According to Dalton's law of partial pressures, the total pressure p in the vessel will then be

$$p = p_{air} + p_v$$

so that if p_{air} is known, the vapor pressure of the liquid can be determined by measuring p.

In this experiment you will use this technique to measure the vapor pressure of tetrachloromethane (CCl_4) between 0°C and 40°C.

REFERENCE   Cromer: *Physics for the Life Sciences*, Sec. 8.5 and Experiment 10.

PROCEDURE

The apparatus consists of a glass bulb connected by a rubber tube to one arm of a flexible-tube manometer. A short tube is connected to the rubber tube by a T-connector (Fig. 11.1). The stopcock connecting the bulb to the manometer is open. *Under no circumstances must you touch the stopcock.* Before each measurement, the manometer arms are adjusted until the mercury level in the bulb-side arm is even with the bottom of the stopcock. In this way, the volume of gas is kept constant while the temperature and pressure of the gas in the bulb are varied.

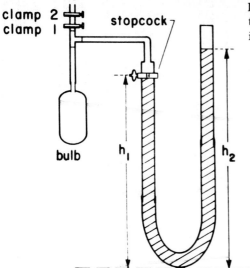

FIGURE 11.1 Flexible-tube manometer connected to a glass bulb by a rubber tube to which a short tube is connected by a T-connector.

Before you begin, lower the bulb-side arm as far as it will go, being careful not to break the glass stem. Be sure to lower the bulb-free arm at the same time, so that the mercury does not rise above the bottom of the stopcock.

Proceed as follows:

1 Remove the clamps from the short tube attached to the T-connector. Adjust the manometer until the mercury in both arms is even with the bottom of the stopcock. This assures that there is a definite volume of air in the bulb at atmospheric pressure. Measure and record the heights h_1 and h_2 of the mercury columns in the two arms, and the temperature of the room.

2 Attach clamp 1 to the short tube, just above the T-connector. Be sure the clamp is tight. Then put an eye-dropper full of CCl_4 into the tube, attach clamp 2 about 2 inches above clamp 1, and remove clamp 1. This procedure enables you to introduce the liquid into the tube without any air escaping. Shake the bulb gently, so that some of the liquid coats the side of the bulb.

REMARK Handle CCl_4 with care because its vapor is hazardous. Work only in a well-ventilated room and avoid breathing CCl_4 vapor directly.

3 As the CCl_4 evaporates, the pressure in the bulb increases. Adjust the bulb-free arm of the manometer so that the mercury level in the bulb-side arm is even with the bottom of the stopcock. Wait several minutes before taking a final reading to be sure that equilibrium has been established in the bulb. Record the temperature and the heights of the mercury columns.

4 Lower the bulb-free arm about 15 cm. Place the bulb in a bucket of ice and water, and wait five minutes for equilibrium to be established. Then adjust the bulb-free arm until the mercury is even with the bottom of the stopcock. Record the data on the data table.

REMARK It is important to lower the bulb-free arm before the bulb is placed in ice water. Why?

5 Take measurements with the bulb in water baths with temperatures around 10°, 20°, 30°, and 40°C. (These baths can be made by mixing ice, and cold and hot tap water in different proportions.) Shake the bulb before each reading.

6 Measure atmospheric pressure with the wall barometer.

ANALYSIS

Calculate the absolute pressure

$$p = \bar{p} + p_0$$

in the bulb at each temperature from your measurements of the gauge pressure $\bar{p}$ and atmospheric pressure p_0.

You will recall that according to Dalton's law of partial pressures, the absolute pressure p is the sum of the partial pressures p_{air} and p_v of the air and the CCl_4 vapor in the bulb

$$p = p_{air} + p_v$$

The partial pressure of air is proportional to the absolute temperature

$$p_{air} = \frac{nR}{V} T = KT$$

At room temperature T_r, the air in the bulb was at atmospheric pressure, so

$$p_0 = KT_r$$

or

$$K = \frac{p_0}{T_r}$$

Thus, the partial pressure of the air in the bulb at any temperature T is

$$p_{air} = \frac{T}{T_r} p_0$$

Calculate p_{air} at each temperature from this last equation. Subtract p_{air} from the measured value of p to find the vapor pressure p_v.

Plot p_v against temperature, and connect the points with a smooth curve.

Name _____

Date _____

REPORT SHEET
EXPERIMENT 11   VAPOR PRESSURE

DATA

Barometric pressure, p_0 = _____

Gas	Temperature t (°C)	Height h_1 (cm)	Height h_2 (cm)	Gauge pressure $\bar{p}$ (mm Hg)	Absolute pressure $p = \bar{p}+p_0$ (mm Hg)	Partial pressure of air p_{air} (mm Hg)	Vapor pressure of CCl_4 p_v (mm Hg)
Air only							
Air + CCl_4							

ANALYSIS

Calculate p and p_{air} and complete the above table to find p_v. Plot p_v against temperature and draw a smooth curve through the points.

QUESTIONS

1. The pressure of the vapor does not increase linearly with temperature, even though the vapor obeys the ideal gas law ($PV = nRT$). Explain this apparent paradox.

2. The vapor pressure of CCl_4 is 317 mm Hg at 50°C. What would be the pressure in the bulb if it were heated to 50°C? Why is it dangerous to do this?

3. When a person breathes air at a pressure of 1 atm, the partial pressure of H_2O in the air in his lungs (alveolar air) is 47 mm Hg. Assuming that alveolar air is in equilibrium with the moisture in the lungs, what is the partial pressure of H_2O in the alveolar air of a diver breathing air at a pressure of 2 atm?

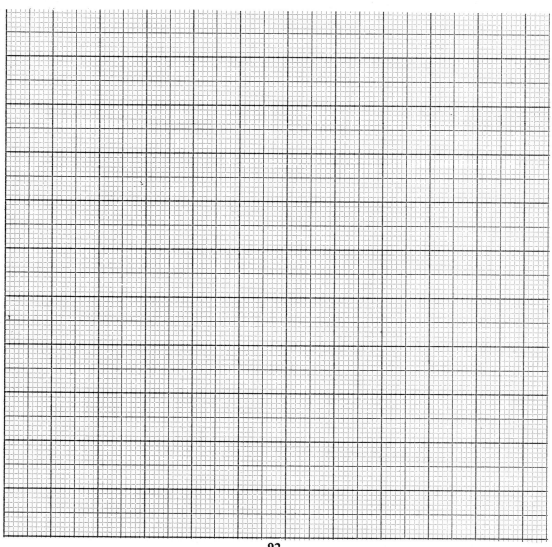

EXPERIMENT 12   SURFACE TENSION

GOALS

1 To demonstrate that the pressure in a bubble is inversely proportional to the radius of the bubble.

2 To demonstrate that the height to which a liquid rises in a capillary is inversely proportional to the radius of the capillary.

3 To measure the surface tension of water using the bubble method and the capillary-action method.

4 To demonstrate the effect of a surfactant on the surface tension of water.

EQUIPMENT

Surface-tension apparatus
Glass capillaries (0.25, 0.50, and 1.00 mm)
Beaker
Plastic ruler
Detergent

INTRODUCTION

A molecule in the interior of a liquid is attracted by all the surrounding molecules, so that the total force on it tends to be zero. A molecule near the surface of the liquid, however, is attracted only by molecules on the liquid side of the surface, so there is a net surface force on the molecule which tends to pull it into the interior of the liquid. Because of this surface force, a liquid tries to make every molecule an interior molecule. This, of course, is impossible, so the liquid does the next best thing; it makes its surface as small as possible.

As a consequence of this tendency, the surface of a liquid acts like a piece of stretched rubber. The surface is under tension and exerts a force perpendicular to any line drawn on the surface (Fig. 12.1) The magnitude F of the force exerted on a line of length l is

$$F = \gamma l$$

where the force per unit length γ is a constant, called the *surface tension*. (Surface tension is somewhat analogous to pressure, which is the the force per unit area in the volume of a fluid.)

One method used to measure the surface tension of a liquid is to dip a capillary tube into the liquid and to increase the pressure in the tube until air bubbles start to appear. The surface-tension apparatus used to measure the pressure in the capillary is a water manometer like the one shown in Fig. 12.2. The pressure in the capillary is greatest when the bubble has the hemispherical shape shown in Fig. 12.3. Any attempt to increase the pressure further results in the release of free bubbles.

To find the relation between the maximum (gauge) pressure $\bar{p}$ in the capillary and the surface tension of the liquid, we calculate the forces on the hemispherical bubble as shown in Fig. 12.3. Because of the excess pressure $\bar{p}$ in the capillary, there is the downward force of magnitude

$$F_1 = \bar{p}A = \bar{p}\pi r^2$$

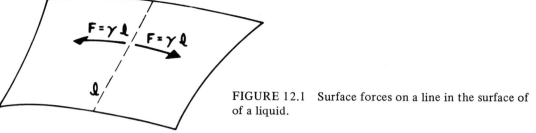

FIGURE 12.1 Surface forces on a line in the surface of of a liquid.

FIGURE 12.2 Surface-tension apparatus. [*Central Scientific Co.*

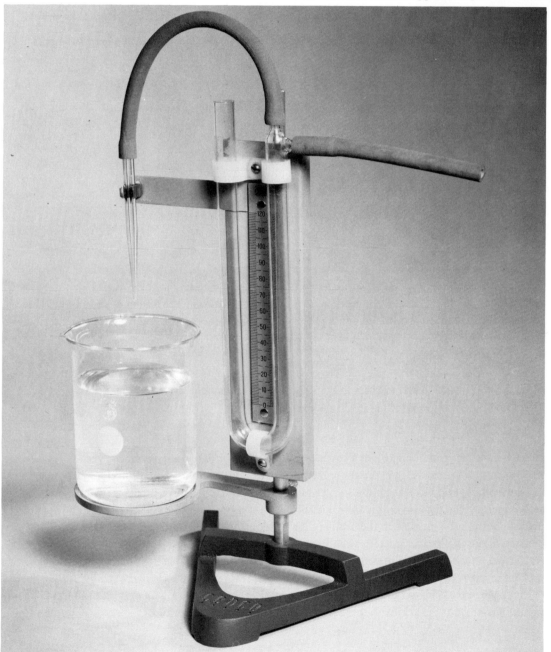

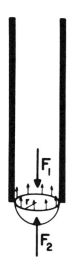

FIGURE 12.3 Forces on a hemispherical bubble.

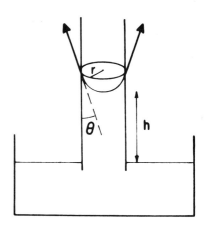

FIGURE 12.4 Forces on a liquid column in a capillary.

on the bubble, where r is the radius of the capillary. The surface tension exerts an upward force along the circumference of the bubble of magnitude

$$F_2 = \gamma 2\pi r$$

At equilibrium these forces are equal and opposite, so

$$\gamma 2\pi r = \overline{p}\pi r^2$$

or

$$\gamma = \tfrac{1}{2}\overline{p}r \qquad\qquad 1$$

Since γ is a constant for a given liquid, this last relation implies that the pressure required to form a bubble decreases as the radius of the bubble increases. In this experiment we shall study this remarkable fact.

Another method used to measure the surface tension of a liquid is to dip a capillary tube into the liquid and measure the height h to which the liquid rises in the tube. A liquid rises in a wet capillary because the liquid clinging to the walls of the capillary exerts an upward force (*capillary action*) on the column of liquid. Figure 12.4 shows that the magnitude of this force is

$$F_2 = \gamma 2\pi r \cos\theta$$

where θ is the angle between the surface of the liquid and the wall of the capillary. Gravity exerts the downward force

$$F_1 = mg = \rho V g = \rho \pi r^2 h g$$

on a column of height h. The liquid rises in the capillary until F_1 is equal to F_2. Thus, at equilibrium

$$\gamma 2\pi r \cos\theta = \rho \pi r^2 h g$$

or

$$\gamma \cos\theta = \tfrac{1}{2}\rho g h r \qquad\qquad 2$$

For water in a clean glass capillary, θ is close enough to $0°$ that $\cos\theta$ can be set equal to 1 in Eq. 2. However, if there is any grease in the tube, θ can be large. Since it is difficult to measure θ directly, in this experiment capillary action will be used to measure $\gamma \cos\theta$, and the result will be compared to the value of γ obtained using the bubble method.

REFERENCE Cromer: *Physics for the Life Sciences*, Sec. 9.3.

Part I Surface Tension of Water

PROCEDURE

Proceed using both the bubble and the capillary-action method.

BUBBLE METHOD You are supplied with three capillaries, with diameters of about 0.25, 0.50, and 1.00 mm. (The exact radius of each capillary is indicated on a tag attached to it.) Connect the 0.25-mm capillary to one arm of the manometer (Fig. 12.2), clamp the capillary in place and adjust it so its tip just touches the surface of the water in the cup. Put a clamp on the second piece of rubber tubing that is attached to the manometer, and increase the pressure in the capillary by rolling the rubber tube around the clamp. Measure the maximum height h' reached by the water in the left-hand arm of the manometer. The height difference Δh between the two arms of the manometer is

$$\Delta h = 2(h' - h_0)$$

where h_0 is the height when the water level is the same in both arms. Repeat with the other two capillaries.

CAPILLARY-ACTION METHOD Place a capillary in a beaker of water and measure with a plastic ruler the height h that the water rises in the capillary. The height is measured from the surface of the water. Be sure that there are no isolated drops of water trapped in the capillary. Repeat with the other capillaries.

ANALYSIS

Analyze the results of the bubble and capillary-action methods as follows:

BUBBLE METHOD The gauge pressure $\bar{p}$ in the capillary is related to the height difference Δh by

$$\bar{p} = \rho g \Delta h$$

Substituting this into Eq. 1 we get

$$\gamma = \tfrac{1}{2} \rho g \Delta h r$$

or

$$\Delta h = \frac{\gamma}{\tfrac{1}{2}\rho g} \frac{1}{r}$$

Thus, a plot of Δh against $1/r$ should be a straight line with a slope

$$s = \frac{\gamma}{\tfrac{1}{2}\rho g} = \frac{\gamma}{4900 \text{ N/m}^3}$$

Plot your measured values of Δh against $1/r$, and calculate the surface tension of water from the slope of the straight line drawn through the points. Be sure that Δh and r are in meters.

CAPILLARY-ACTION METHOD From Eq. 2 we have

$$h = \frac{\gamma \cos \theta}{\frac{1}{2} \rho g} \frac{1}{r}$$

so that a plot of h against $1/r$ should be a straight line with a slope

$$s = \frac{\gamma \cos \theta}{\frac{1}{2} \rho g} = \frac{\gamma \cos \theta}{4900 \text{ N/m}^3}$$

Plot your measured values of h against $1/r$, and calculate $\gamma \cos \theta$ from the slope of the straight line drawn through the points. Use the result of the bubble method to find $\cos \theta$.

Part II Surfactants

PROCEDURE

A surfactant is any substance that when added to water in a small quantity greatly reduces the surface tension. Add 5 or 10 ml of some surfactant, such as detergent or "wetting solution", to a beaker of water. Repeat the measurements of Part I using both methods to determine the surface tension for this solution.

ANALYSIS

Repeat the analysis of Part I, and compare the values of the surface tension found in the two cases.

Name _____

Date _____

REPORT SHEET
EXPERIMENT 12   SURFACE TENSION

Part I   Surface Tension of Water

DATA

Capillary	Radius of the capillary r (m)	$1/r$ (m^{-1})	Bubble method h (m)	h_0 (m)	Δh (m)	Capillary-action method h (m)
0.25 mm						
0.50 mm						
1.00 mm						

ANALYSIS

Analyze the results of the bubble and capillary-action methods as follows;

BUBBLE METHOD Calculate Δh and enter on data table. Plot Δh against $1/r$ and determine γ from the slope of the straight line drawn through the data points.

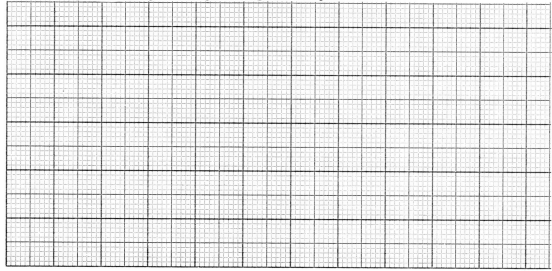

Surface tension of water, γ = _____

CAPILLARY-ACTION METHOD Plot h against $1/r$ and determine $\gamma \cos \theta$ from the slope of the straight line drawn through the data points.

$\gamma \cos \theta = $ _____

$\cos \theta = $ _____

QUESTIONS

1. The surface tension of acetone (density = 792 kg/m³) is measured using both the bubble method and the capillary action method. With a capillary of radius 0.15 mm, the data are Δh = 32 mm and h = 39 mm. Calculate γ and $\cos \theta$.

2. Which of the two methods of measuring the surface tension of a liquid does not require knowledge of the density of the liquid?

3. A leg of a water spider makes a hemispherical depression of radius 0.3 cm in the surface of the water. (a) What is the force which the leg exerts on the water? (b) What is the mass of the spider?

Part II   Surfactants

DATA

Capillary	Radius of the capillary		Bubble method			Capillary-action method
	r (m)	$1/r$ (m⁻¹)	h (m)	h_0 (m)	Δh (m)	h (m)
0.25 mm						
0.50 mm						
1.00 mm						

ANALYSIS

Calculate Δh and enter on the data table. Plot Δh and h against $1/r$ on the graphs used in Part I. Determine γ, $\gamma \cos \theta$, and $\cos \theta$ from the slopes of the straight lines drawn through these points.

$\gamma =$ _____

$\gamma \cos \theta =$ _____

$\cos \theta =$ _____

QUESTIONS

1 The force between a surfactant molecule and a H_2O molecule is weaker than the force between the H_2O molecules. Explain how this results in a greater concentration of surfactant molecules on the surface of the liquid than in the interior.

2 Describe a procedure for determining the concentration of a surfactant in water by measurements of surface tension.

wave phenomena

EXPERIMENT 13   STANDING WAVES

GOALS

1. To observe standing waves on a string.
2. To demonstrate that the wavelength of a standing wave on a string is proportional to the square root of the tension of the string.
3. To measure the speed of sound by establishing standing waves in an air column.

EQUIPMENT

Fixed-frequency vibrator
Cotton string
Pulley
Shot can and shot

Balance (up to 2 kg)
Meterstick
Sound-wave resonance apparatus with scale
4 tuning forks ranging from 256 Hz to 2048 Hz

INTRODUCTION

A standing wave is a wave that remains in a fixed position rather than traveling through a medium. It can exist in any medium that has a boundary beyond which a wave cannot propagate, because when a wave traveling in the medium comes to such a boundary it is reflected back into the direction from which it came. Consequently, when a train of traveling waves impinges on a fixed boundary, the reflected wave superposes with the incident wave to produce a standing wave pattern in the medium. Sinusoidal transverse waves on a string are illustrated in Fig. 13.1, where the result of the superposition of the reflected and incident wavetrains is shown. This resultant is called a *standing wave* because the points of zero amplitude (the *nodes*) are stationary. A nodal point is always at rest. As can be seen from Fig. 13.1, a given node is one-half wavelength away from its neighbors. The string vibrates only in the regions between the nodes, and the midpoint of such a region, where the amplitude reaches its maximum value, is called an *antinode*.

Consider a guitar string which is fixed at both ends. What determines the frequency at which this string can vibrate? It can be seen that both endpoints must be nodes in the standing-wave pattern, since there can be no displacement there. The longest wave which is able to satisfy these "boundary conditions" is one without any nodes in between the end points (only one antinode). If the length of the string is L, then the wavelength of this standing wave is $\lambda_1 = 2L$. The next longest wave to satisfy these conditions is a standing wave with only one node between the end points. It has a wavelength given by $\lambda_2 = 2L/2 = L$. The next longest wave will have a wavelength of $\lambda_3 = 2L/3$, and so on. In general, all wavelengths given by

$$\lambda_n = \frac{2L}{n} \quad (n = 1, 2, 3, \ldots) \qquad\qquad 1$$

can exist on the guitar string.

The frequency of any wave is obtained by dividing the speed of the wave v by the wavelength.

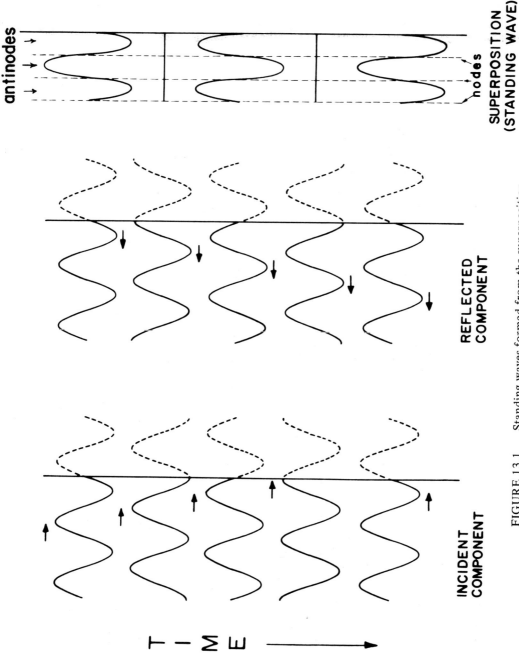

FIGURE 13.1 Standing waves formed from the superposition of a sinusoidal wave traveling to the right and its reflected wave traveling to the left.

This results in the following possible frequencies of vibration for a transverse standing wave on a string with fixed ends

$$f_n = \frac{v}{\lambda_n} = \frac{nv}{2L} \qquad \qquad 2$$

The condition of the boundary in a given situation is very important in determining the allowed frequencies of vibration. It is possible, for example, that one end of the string is an antinode. This leads to a different set of allowed frequencies.

In this experiment, you will investigate standing wave patterns produced with transverse waves on a string and with longitudinal sound waves in air. Your instructor will explain the distinction between these two types of wave motion.

REFERENCE Cromer: *Physics for the Life Sciences*, Chap. 12, Sec. 13.1 and 13.4.

Part I   Transverse Waves on A String

PROCEDURE

The apparatus consists of an electric **vib**rator with fixed frequency, which vibrates one end of a stretched cotton string. The other end of the string passes over a pulley and is attached to a shot can. The tension of the string, and thus the speed of the wave, is adjusted by varying the amount of shot in the can. Start the vibrator and add shot to the can until a standing-wave pattern with at least six half-wave segments is obtained. A standing-wave pattern with an integral number of half-wave segments will be apparent by the relatively large amplitude at the antinodes.

Measure the distance between nodes and weigh the shot can. Continue adding more shot until one less half-wave segment is produced. Again measure the distance between nodes and weigh the shot can. Repeat this procedure until only three half-wave segments are produced.

ANALYSIS

Plot the square of the measured wavelength values against the string tension. According to theory, the speed of transverse waves on a string is

$$v = \sqrt{\frac{TL}{M}} \qquad \qquad 3$$

where T is the tension, L is the length, and M is the mass of the string. The wavelength λ_n is related to the fixed frequency f by

$$\lambda_n = \frac{v}{f} = \frac{1}{f}\sqrt{\frac{TL}{M}} \qquad \qquad 4$$

or

$$\lambda_n^2 = \frac{1}{f^2}\left(\frac{TL}{M}\right) = \left(\frac{1}{\mu f^2}\right) T$$

where $\mu = M/L$ is the *linear mass density* of the string. The value of μ is marked on the apparatus. The slope s of a plot of λ_n^2 against T is therefore equal to $1/\mu f^2$, and so f is given by:

$$f = \sqrt{\frac{1}{\mu s}} \qquad \qquad 5$$

Draw the best straight line through the data and obtain the slope. Use Eq. 5 to find the frequency of the vibrator and compare your value with the fixed-frequency value posted on the apparatus.

Part II   Speed of Sound Waves in Air

PROCEDURE

The sound-wave resonance apparatus consists of a long glass tube that is connected to a movable reservoir by means of a rubber siphon (Fig. 13.2). The water level in the tube and, thus, the length of the air column is controlled by the position of the reservoir. The speed of the sound waves is constant in this part of the experiment because the air temperature is not changed. Instead, the length of the air column is varied in order to create the resonant condition for standing sound waves along the length of the column. The desired column length is obtained by changing the water level which forms the closed end of the column. The surface of the water, the closed end, corresponds to a node in the standing-wave pattern of the air displacement, while the open end is an antinode.

Start with the water level as high as possible. Strike the 256-Hz tuning fork and hold it over the open end of the tube. Lower the water level slowly until a maximum sound intensity is heard and record the scale value. Continue to lower the level of the water until a second maximum is detected. Be sure that the maxima you produce are at the fundamental frequency of the tuning fork and not at one of the harmonics.

The distance between the positions of the water column for two adjacent maximum sound intensities is equal to ½λ at the frequency used. Repeat the measurements for three higher-frequency tuning forks.

ANALYSIS

Plot the wavelength values λ obtained against $1/f$, and draw the best straight line through the data. Determine the speed of sound in air from the slope of this plot. Compare your value for the speed of sound in air with the standard value.

REMARK   This portion of the experiment can be presented by the instructor in the form of a demonstration if it appears that there is not sufficient time for the class to complete it.

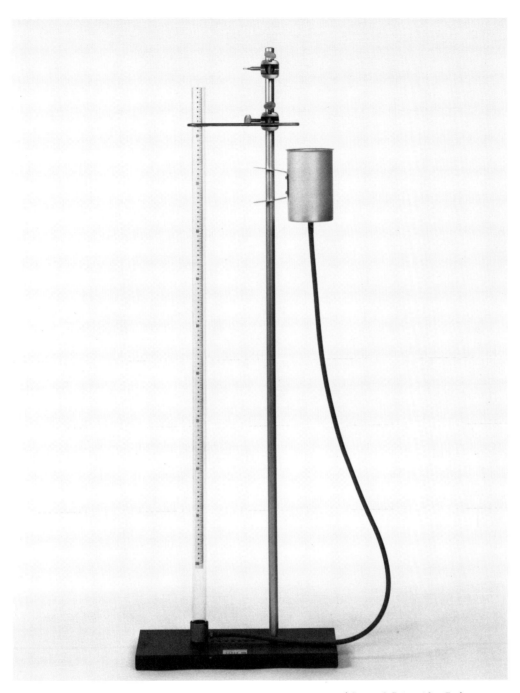

FIGURE 13.2 Sound-wave resonance apparatus (*Central Scientific Co.*).

Name_____

Date_____

REPORT SHEET
EXPERIMENT 13   STANDING WAVES

Part I   Transverse Waves on A String

DATA

No. of half-wave segments	Distance between nodes (m)	Wavelengths λ (m)	Wavelengths λ^2 (m^2)	Tension (N)

ANALYSIS

Plot λ^2 against T below, draw the best straight line through the data, and find the slope. Record the linear mass density of the string and calculate the frequency, f.

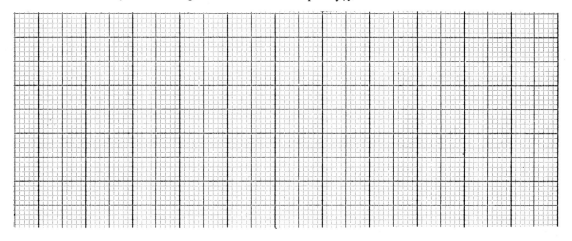

Slope, s = _____ m^2/N

Linear mass density, μ = _____ kg/m

$f = \sqrt{\dfrac{1}{\mu s}}$ = _____ Hz

QUESTIONS

1. Calculate the numerical value of the wave speed for the highest and the lowest tension measured.
2. Would the wave speed in a lead (Pb) wire of the same size and under the same tension as the string used here be greater or less than you measured?
3. By how much does the speed of a wave change if the string tension is (*a*) increased by a factor of nine, and (*b*) decreased by a factor of four?

Part II Speed of Sound Waves in Air

DATA

Fork frequency (Hz)	$1/f$ (Hz^{-1})	Distance between nodes (m)	Wavelength λ (m)

ANALYSIS

Plot λ against $1/f$ on the graph below. Draw the best straight line through the data and find the slope. The slope is equal to the speed of sound.

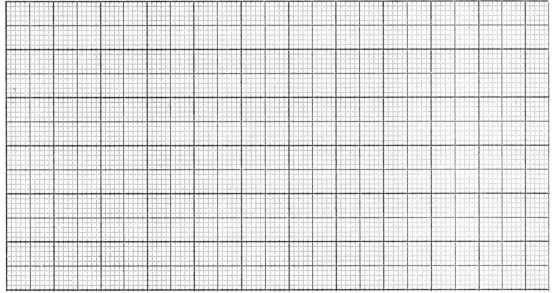

Slope, s = Speed of sound = _____ m/s

QUESTIONS

1 Draw the standing-wave pattern of the displacement of the air column which corresponds to (*a*) the first maximum, and (*b*) the second maximum in the intensity.

2 Draw the corresponding patterns for the pressure in the air column.

3 Compare the speed of sound to the speed of waves on a string and to the speed of light.

EXPERIMENT 14   INTERFERENCE AND DIFFRACTION

GOALS

1 To observe the interference pattern from double and multiple slits.

2 To measure the spectrum of mercury using a diffraction grating.

EQUIPMENT

Slit slide (one through six slits)
Red or blue filter glass
Diffraction grating (15,000 lines per inch)
Optical bench
Sodium lamp
Mercury spectrum tube with power supply
Metersticks
Spectrum wall chart

INTRODUCTION

The principle of superposition states that, at a point at which two waves coexist, the magnitude of the total disturbance is equal to the sum of the disturbances that each wave would produce by itself. Since both positive and negative disturbances can occur, it is possible for the positive disturbance of one wave to cancel the negative disturbance of the other, so that the two waves together can produce a net disturbance of zero at a point where either wave alone would have produced a disturbance. This phenomenon, called *destructive interference*, is peculiar to wave motion, and it is the observation of destructive interference in light that demonstrates that light is a wave. Similarly, if two positive or two negative disturbances exist at the same point, the resulting intensity is greater than the intensity of either disturbance by itself. This phenomenon is called *constructive interference*.

Another general feature of wave motion is the tendency of a wave to spread out after it passes through a narrow opening. This phenomenon is called *diffraction*. In the case of sound waves, this phenomenon enables one to hear the sound from a loudspeaker through a doorway or window, even though the loudspeaker cannot be seen. Light, on the other hand, is usually thought of as traveling along straight lines called rays. However, if a very small opening or slit in an otherwise opaque screen is illuminated with monochromatic light, a distinct pattern of light and dark fringes can be observed, instead of the simple geometrical shadow one would expect on the basis of the ray model.

In this experiment, you will study these wave properties and see how they may be used to measure the wavelength of monochromatic light.

REFERENCE   Cromer: *Physics for the Life Sciences*, Sec. 14.2.

Part I   Interference and Diffraction from Slits

Consider a pair of very narrow slits illuminated by a point source of monochromatic light, which is light of a single wavelength λ. If the width of the individual slits is negligible compared to λ (which is extremely difficult to arrange in practice), the light will radiate uniformly in all directions from

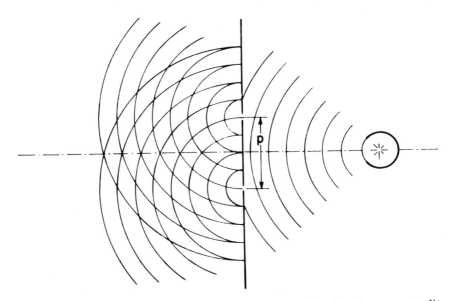

FIGURE 14.1   Interference of waves emerging from two narrow slits.

each slit (Fig. 14.1), so that each slit acts as a line-source of light. Furthermore, since the waves emerging from each slit are part of the single wave produced by the point source, these two waves are *correlated* in the sense that minimum or maximum disturbance is produced simultaneously at each slit.

The two slits may thus be considered to be sources of waves that are correlated in time. The successive wavefronts, which emanate from these two secondary sources, are shown in Fig. 14.1. The circular lines can be regarded as maxima or peaks in the wave and the midpoint of the empty regions in between are the minima or troughs. In order to determine what the interference pattern on a projection screen would be, we use the fact that when a peak from one wavelet superposes on a peak from the other (or a trough with another trough) they add constructively, and a maximum or bright fringe is formed in the total pattern. Conversely, when a peak superposes with a trough, they cancel each other and a dark fringe is formed. Thus, if a screen is placed in position beyond the slits, a pattern of light and dark fringes will appear at the places where the wave pattern produces constructive and destructive interference, respectively. One can easily construct the resulting interference pattern, which to a good approximation consists of a central bright fringe surrounded by a series of equally-spaced fringes that are separated by the distance Δx given by

$$\Delta x = \frac{\lambda}{d} D \qquad\qquad 1$$

where d is the distance between the two slits and D is the distance between the slits and the screen.

PROCEDURE

Examine the sets of slits furnished you, starting with the double slit and working your way up to the sextuple slit. Hold the glass plate close to your eye and look through the slits at the incadescent light. Use a red or blue filter glass in order to make the light monochromatic. Sketch your observations *carefully* on the report sheet.

The double-slit pattern is not exactly what we had predicted. This deviation is due to the *diffraction* of light at each slit. If a slit has a width that is not negligible compared to the wavelength of light, then every point between its edges must be regarded as a separate source of light waves. This

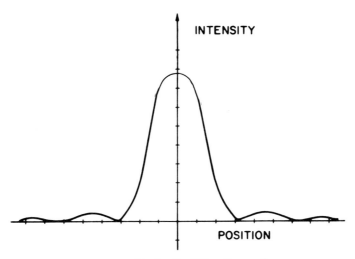

FIGURE 14.2 Single-slit diffraction pattern.

infinite number of slit sources produces its own special kind of interference pattern, illustrated in Fig. 14.2. The relevant formula giving the fringe separation in this pattern contains, of course, the slit width s as the comparative dimension. The distance of separation for the dark diffraction fringes (except for the two on either side of the central maximum) is given by

$$\Delta x = \left(\frac{\lambda}{s}\right) D \qquad \qquad 2$$

The width of the central maximum is twice this distance.

ANALYSIS

Examine the *single* slit and see if you can observe the diffraction pattern. Make an accurate sketch on the report sheet. Identify and indicate the diffraction effects in the sketches of the multiple slit patterns.

Part II   Diffraction Grating

A plate with a large number of closely spaced slits is called a *diffraction grating*. As you may have deduced in Part I, the larger the number of slits, the sharper the main bright fringes become. It can be shown that the fringe pattern obeys the following relation

$$\sin \theta = \frac{n\lambda}{d} \qquad (n = 0, 1, 2, \ldots) \qquad \qquad 3$$

where θ is the angular position of the nth order fringe relative to the center line (Fig. 14.3), and d is the distance between adjacent slits. Diffraction gratings are usually specified in terms of the number of lines (slits) per unit length. This number must be inverted to obtain d.

PROCEDURE

In this part of the experiment you will measure the wavelengths of several of the prominent lines in the spectrum of mercury.

Mount the adjustable slit in the center of a meterstick or optical bench, and mount the diffraction grating in a holder about 2 m from the slit (Fig. 14.3). Measure the distance *a* between the grating and the slit.

The grating must be calibrated, because the nominal value given for the number of slits per unit length is only approximate. This is done using the known wavelength of the yellow light of sodium vapor lamp. Turn on the sodium lamp and place it behind the slit. With the 15000-lines-per-inch diffraction grating in position, view the pattern through the grating (Fig. 14.3). Measure and record the positions of the first order yellow lines and determine the distance *b* between them. You will be able to read the scale on the meterstick or optical bench (a little front illumination sometimes helps).

Replace the sodium lamp with the mercury lamp. Measure and record the values of *b* corresponding to each spectral line you observe.

ANALYSIS

From Eq. 3, we have

$$d = \frac{n\lambda}{\sin\theta} \qquad 4$$

Calculate *d* from this relation, given that the wavelength λ of the yellow light of sodium is 589.2 nm (1 nm = 10^{-9} m) and that $\sin\theta$ is found from the right triangle in Fig. 14.3 to be

$$\sin\theta = \frac{\text{opposite side}}{\text{hypotenuse}} = \frac{\tfrac{1}{2}b}{\sqrt{a^2 + (\tfrac{1}{2}b)^2}} \qquad 5$$

Use the value of *d* obtained above to compute the wavelength of each mercury line. Some of the prominent mercury lines are 404.7, 407.7, 435.8, 491.6, 546.0, 577.0, and 579.0 nm. A spectrum wall chart is also available for comparison.

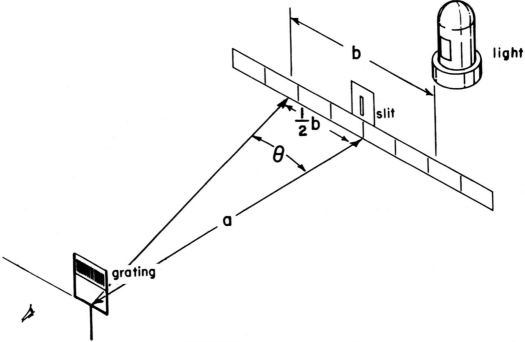

FIGURE 14.3 Arrangement of the apparatus for measurements with the diffraction grating.

Name_____

Date_____

REPORT SHEET
EXPERIMENT 14   INTERFERENCE AND DIFFRACTION OF LIGHT

Part I   Interference and Diffraction From Slits

DATA AND ANALYSIS

Sketch below the interference patterns of the slit sets. Identify and indicate the diffraction effects in the sketches.

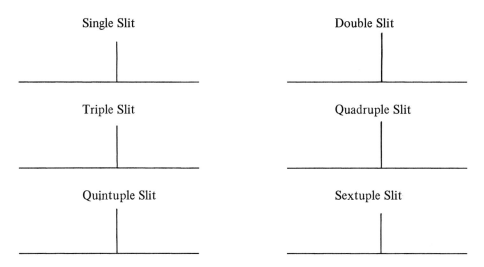

QUESTIONS

1. What happens as the number of slits is increased? Be explicit.
2. For the double slits, determine which is greater, the slit separation or the slit width, using Eqs. 1 and 2. Examine the pattern carefully.
3. Give an example of the effects of diffraction in everyday life.

Part II   Diffraction Grating

DATA

  Distance, a = _____ cm

  Sodium-lamp data: Distance, b = _____ cm

     Wavelength, λ = 589.2 nm

Mercury-lamp data:

Line	Color	Distance b (m)	½b, (cm)	$\sin\theta = \dfrac{\frac{1}{2}b}{\sqrt{a^2 + (\frac{1}{2}b)^2}}$	$\lambda = d\sin\theta$, (nm)
1					
2					
3					
4					
5					
6					
7					

ANALYSIS

Determine the slit separation d from the sodium-lamp data by completing the following calculations. First, calculate $\sin\theta$ from Eq. 5:

$$\sin\theta = \dfrac{\frac{1}{2}b}{\sqrt{a^2 + (\frac{1}{2}b)^2}} \underline{\hspace{3cm}}$$

Then calculate d from Eq. 4:

$$d = n\left(\dfrac{\lambda}{\sin\theta}\right) = 1 \times \dfrac{589.2 \times 10^{-7}\text{ cm}}{\sin\theta} = \underline{\hspace{3cm}} \text{ cm}$$

Complete the entries in the mercury-lamp table using this value of d.

QUESTIONS

1 Which end of the spectrum (red or violet) experiences the largest displacement along the screen?

2 What would happen to the distance b between the two first-order sodium yellow lines if the number of lines per inch on the grating were doubled?

3 What does the number n in Eq. 3 signify?

EXPERIMENT 15   REFLECTION AND REFRACTION

GOALS

1 To demonstrate the laws of reflection and refraction.

2 To show how the law of reflection determines the location of the virtual image formed by a mirror.

3 To measure the index of refraction of glass.

4 To measure the critical angle for total internal reflection.

EQUIPMENT

Ray box
Plane mirror
Half-dozen hat pins
2' X 2' smooth flat pin board
Sheets 8½" X 11" paper

Protractor
Ruler
Rectangular glass plate with polished edges
Triangular prism

INTRODUCTION

When a beam of light is incident on the boundary between two transparent media, part of the beam is *reflected* from the boundary back into the first medium, and part of the beam proceeds into the second medium. The part that penetrates into the second medium is called the *refracted* wave. In this experiment we shall study the laws of reflection and refraction that relate the angles that the incident, reflected, and refracted beams make with the boundary. These laws are important in order to understand the operation of basic optical elements, such as mirrors, prisms, and lenses, that are used to reflect and focus light in optical instruments.

Narrow beams of light, or rays, can be thought of as emanating outward in all directions from each point of an illuminated object. We see an object because rays coming from the same point are focused by the lens system of the eye onto a single point of the retina, while the rays coming from other points of the object are focused on different points of the retina (Fig. 15.1). Consequently, a real image of the object is formed on the retina. When an optical element, such as a mirror or a piece of glass, is placed between the object and the eye, the rays of light that enter the eye no longer appear to come from where the object is located. Figure 15.2 shows that in the case of the mirror the object appears to be behind the mirror, because the rays of light entering the eye, when traced back to their intersection, appear to come from points behind the mirror. We say that a *virtual* image of the object is formed in back of the mirror, the word virtual indicating that the image is not real.

In this experiment you will study rays reflected from a mirror and refracted by a glass plate, and confirm that the positions of the virtual images formed by these devices are determined by the laws of reflection and refraction applied to the rays viewed by the eye.

REFERENCE   Cromer: *Physics for the Life Sciences*, Sec. 14.3.

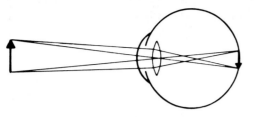

FIGURE 15.1   Image focused on the retina of the eye.

Part I   Reflection

PROCEDURE 1

The apparatus consists of a plane mirror, hat pins, a large pin board, and a ray box which projects several narrow beams of light along a flat surface.

Draw a straight line in the middle of a plain sheet of paper and attach the paper to the pin board with masking tape. Position the mirror directly on this line. Turn on the ray box and direct the rays toward the mirror at about a 45° angle. Trace the path of an incident and reflected ray by making pencil marks at two points along the length of each ray. Connect these points with a straight edge. Repeat this procedure for two other incident angles.

ANALYSIS 1

With a protractor, construct a line perpendicular to the plane of the mirror at the point where each incident ray intersects the mirror's surface. This line is called a *normal*. The angle θ_1 between the normal and the incident ray is called the *incident* angle, and the angle θ_1' between the normal and the reflected ray is called the *reflected* angle (Fig. 15.3). Measure and record each pair of incident and reflected angles on the report sheet.

The law of reflection states that *the angle of reflection is equal to the angle of incidence.* Check whether your data is consistent with this law.

FIGURE 15.2   Virtual image formed by a plane mirror.

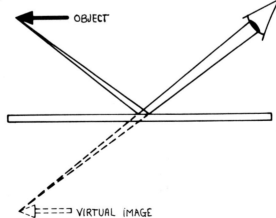

FIGURE 15.3   Incident and reflected rays.

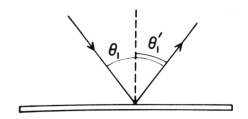

PROCEDURE 2

On a fresh sheet of paper, draw an arrow AB about 2 cm long, at a position which is about 4 cm from the edge of the paper. Directly across the middle of the sheet, and exactly parallel to the arrow, draw a straight line and place the plane mirror over it. Place a pin at point B. The apparent position of the virtual image of point B can be located by the following method. Put two pins at points B_1 and B_2 so that the image of the pin at B appears to be lined up with the pins at B_1 and B_2. Put two other pins at points B_3 and B_4, so that the image of the pin at point B appears to line up with the pins at B_3 and B_4 (Fig. 15.4). Remove the pins and mark their positions carefully. Remove the mirror and extend the lines B_1B_2 and B_3B_4 until they intersect. This point will be the position B' of the image of the pin.

Replace the mirror exactly as before. Place a pin at A and, using the above technique, find the position of its image and call it A'. Since the image is a straight line just like the object, the line $A'B'$ is the image of the arrow AB as seen in the mirror.

ANALYSIS 2

The position and size of the image is determined entirely by the law of reflection, since no light passes through the mirror. To confirm this draw a straight line from point B to the point where the line B_1B_2 meets the mirror. Draw the normal to the mirror at this point. Measure the incident and reflected angles to confirm that the law of reflection is obeyed. Next, measure the distance s between the object and the mirror and the distance s' between the image and the mirror. Measure the length h of the object and the length h' of its image.

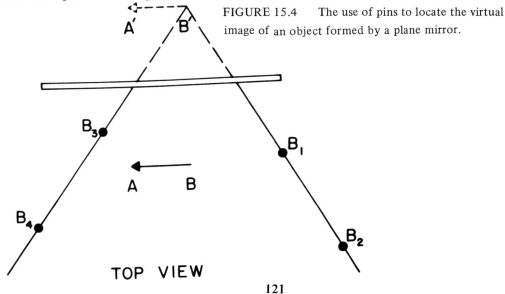

FIGURE 15.4   The use of pins to locate the virtual image of an object formed by a plane mirror.

FIGURE 15.5 Incident, reflected, and refracted rays.

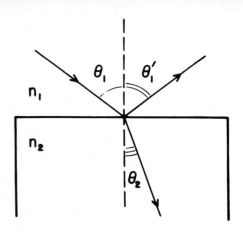

Part II Refraction

When light passes through the boundary between two transparent media, its direction of propagation changes in accordance with *Snell's law*, which relates the angle of incidence θ_1, the angle of refraction θ_2, and the indices of refraction n_1 and n_2 of the two media as follows

$$n_1 \sin \theta_1 = n_2 \sin \theta_2 \qquad \qquad 1$$

The angles θ_1 and θ_2 are both measured with respect to the normal, as shown in Fig. 15.5.

In this experiment, medium 1 is air so that, to a good approximation, $n_1 = 1$.

PROCEDURE

Place a rectangular glass plate at the center of a sheet of plain white paper, which is attached to the pin board. Outline the plate with a sharp pencil. Turn on the ray box and direct it toward the plate so that the rays make an angle with the normal to the air-glass boundary. Trace the entrant and emergent rays on the paper by using the same procedure as before. Remove the plate, draw the lines, and label them to correspond to Fig. 15.6. Label and measure the angles *a, b, c, d* to the nearest 0.5° with a protractor. Record the values on the report sheet. Repeat this procedure for three other angles of incidence.

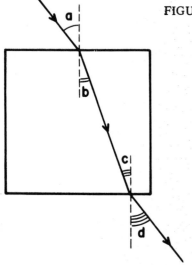

FIGURE 15.6 Rays passing through a rectangular piece of glass.

ANALYSIS

Find the sine of each angle in the four sets of measurements that were taken. (You should be able to convince yourself that angles a and d, and angles b and c are equal, respectively). Plot the sine of the incident angle against the sine of the refraction angle in each case. Draw the best straight line through the data and find the slope.

According to Snell's law

$$\sin \theta_1 = \frac{n_{glass}}{n_1} \sin \theta_2 \qquad\qquad 2$$

where θ_1 and θ_2 are the incident and refracted angles, respectively. The slope obtained from the plot, therefore, is equal to the index of refraction of glass, since $n_1 = 1$.

Part III   Total Internal Reflection

An interesting phenomenon occurs when a light ray in one medium (such as glass) is incident on the boundary of another medium (such as air) that has a *lower* index of refraction. From Snell's law the angle of refraction θ_2 is related to the angle of incidence θ_1 by

$$\sin \theta_2 = n_{glass} \sin \theta_1$$

Since n_{glass} is greater than 1, the refraction angle θ_2 will always be greater than the incident angle θ_1. Therefore, as θ_1 is increased, a value of $\sin \theta_1$ will be reached for which $\sin \theta_2 = n_{glass} \sin \theta_1 = 1$, and so the refraction angle is 90°. This value of θ_1 is called the *critical angle for total internal reflection*, because for any angle greater than this there is no refracted ray, and all the light incident on the boundary is reflected back into medium 1.

PROCEDURE

Place a triangular prism at the center of a plain white sheet of paper attached to the pin board. Outline the prism with a sharp pencil. Turn on the ray box and direct it toward one face of the prism. Slowly rotate the ray box in such a way as to increase the refraction angle of the exiting ray. At a point just before this angle becomes equal to 90° record the ray incident on the prism as well as the point at which the ray leaves the prism. (This condition is also accompanied by the phenomenon of *dispersion* in which the white light of the ray is split up into the component colors of the spectrum.)

FIGURE 15.7   Rays passing through a prism.

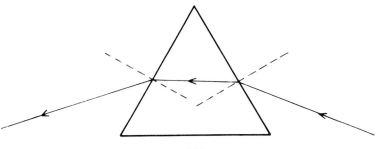

ANALYSIS

Remove the prism and draw the lines representing the path of the ray (Fig. 15.7). At the interface where the ray left the prism, measure the angle of incidence θ_1. Use Snell's law to determine n_{glass} from the condition for total internal reflection, and compare this value with that obtained in Part II. If time permits, it is advisable to also measure the incident and refraction angles at the entrance interface and apply Snell's law there to obtain a doublecheck on the value of the index of refraction. (Bear in mind that all angles are measured from the normal to the interface.)

Name_____

Date_____

REPORT SHEET
EXPERIMENT 15   REFLECTION AND REFRACTION

REMARK The four sheets of paper on which the geometrical constructions and measurements were made must be included in this report.

Part I   Reflection

DATA

The data are the sheets of paper on which the rays were traced. Label them Procedure 1 and Procedure 2, respectively.

ANALYSIS

Analyze the data from Procedures 1 and 2 as follows.

Procedure 1 Measure the incident and reflected angles for each ray and enter in the table.

Incident angle, θ_1	Reflected angle, θ'_1

Procedure 2 Measure the incident and reflected angles of the ray that goes from point B to the point where the line $B_1 B_2$ intersects the mirror.

$\theta_1 =$ _____ $\theta'_1 =$ _____

Measure the object and image distances, s and s'.

$s =$ _____ $s' =$ _____

Measure the lengths h and h' of the object and image.

$h =$ _____ $h' =$ _____

QUESTIONS

1 What is the relation between the image distances and the object distance s' from the plane mirror?
2 Magnification m is defined as the ratio of the image size h' to the object size h. What is the value of the magnification in the case of a plane mirror?
3 Use the law of reflection to determine the position(s) of the image(s) of an object located on the bisector of the angle between two mirrors inclined at 90° to each other, as shown in Fig. 15.8.

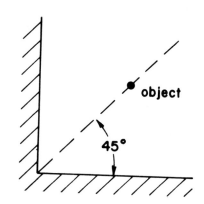

FIGURE 15.8

Part II   Refraction

DATA

Trial No.	a	b	c	d	$\sin \theta_1$	$\sin \theta_2$
1						
2						
3						
4						

ANALYSIS

Calculate $\sin \theta_1$ and $\sin \theta_2$ and complete the data table. Plot $\sin \theta_1$ against $\sin \theta_2$ on the graph below. Draw the best straight line through the data and find the slope s.

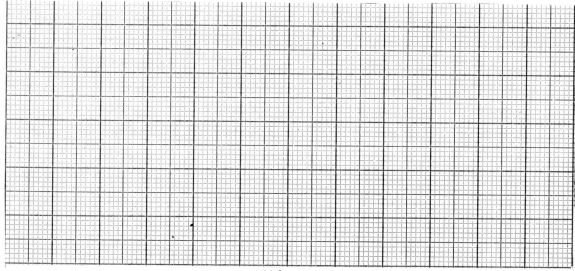

s = _____

= n_{glass}

Value of n_{glass} from textbook = _____

QUESTIONS

1 What is the angle between the rays which enter and the rays which emerge from the glass?
2 Notice that the emergent ray is displaced relative to the entrant ray. For a given angle of incidence in air, what are the two properties of the glass plate that determine the amount of lateral displacement?
3 Explain why a straight stick that is partly immersed in water appears to be bent.

Part III   Total Internal Reflection

DATA

The angle of incidence at the second interface when $\theta_2 = 90°$,

θ_1 = _____

$\sin \theta_1$ = _____

ANALYSIS

Determine the index of refraction of glass.

n_{glass} = _____

QUESTIONS

1 A fish swimming 3 ft below the surface of a perfectly smooth lake will see only a limited region of the surface illuminated. Determine the radius of the illuminated region. (The index of refraction of water is 1.33.)
2 What does the observation that a prism can disperse the spectrum of white light imply about the ray model and Snell's law?

EXPERIMENT 16   LENSES

GOALS

1 To measure the focal length of a simple converging lens.

2 To observe the real image formed by a converging lens.

3 To study the lens formula.

4 To build a telescope.

EQUIPMENT

Optical bench
Converging lenses of long
    and short focal length
Lighted image
Ground-glass screen
Ruler

INTRODUCTION

A lens works by virtue of the refraction of light at its spherically-shaped surfaces. Any spherically-shaped boundary between two media with different indices of refraction will have focusing or defocusing effects on the light passing through it. Each such surface can form an image of a point object. By constructing a series of such surfaces between different media, most commonly air and glass, a vast variety of optical devices can be made.

A simple lens has two spherical surfaces, one on each side of a piece of glass. The centers of curvature of the surfaces of the lens define a line called the *axis* of the lens. Another fundamental quantity used in describing the properties of a lens is the *focal point* or *back focus* F', which is defined as the point at which incident rays parallel to the axis are brought to focus. The *focal length f* is the distance from the center of the lens to the focal point (Fig. 16.1a). An imaginary plane, which passes through the focal point and which is perpendicular to the lens axis, is called the *focal plane.* This plane has the property that *all* sets of incident parallel rays are focused on it.

Because the properties of a thin lens are symmetric about the center, it is convenient to define a second focal point called the *front focus F,* which is located on the side opposite the back focus at a distance f from the center. Rays emanating from this point emerge parallel to the axis after passing through the lens.

All of these terms are illustrated in Fig. 16.1. Incident parallel rays are imaged by the converging lens on the back focal plane (Fig. 16.1a and *b*), and rays emanating from an object located at the front focus F produces parallel rays (Fig. 16.1c).

Figure 16.2 shows a converging lens with an object located a distance s from the center of the lens. The location of the image is determined by tracing three *principal* rays. The first ray is drawn parallel to the axis and is bent so as to pass through the back focus when it goes through the lens. The second ray passes through the *center* of the lens. If the lens is relatively thin, this ray is not bent by it. Finally, the third ray passes through the front focus and is, therefore, rendered parallel to the

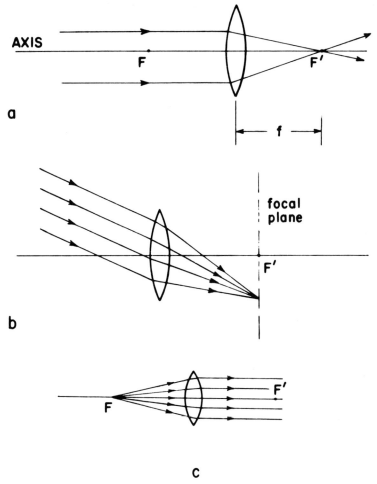

FIGURE 16.1 Rays passing through a converging (short focal length) lens. (*a*) Rays parallel to the axis are brought to a focus at the back focus F'. (*b*) Parallel rays are brought to a focus at a point in the focal plane. (*c*) Rays emanating from the front focus F emerge parallel to the axis.

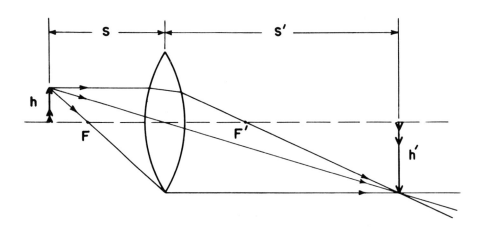

FIGURE 16.2 Principal rays used to locate the image of an object.

axis by the lens. The point where these three rays intersect is the image point corresponding to the object point. In this experiment, you will be studying how lenses work and some of their applications.

REFERENCE   Cromer: *Physics for the Life Sciences*, Sec. 15.1, 15.2, and 15.4.

Part I   Focal Length of the Converging Lens

PROCEDURE

The apparatus consists of an optical bench, several lens holders (large and small), a converging len, a ground-glass screen, a plane mirror, and an illuminated object with holder.

REMARK   Be sure to handle all lenses with the greatest care. If unmounted, take hold of them only by the edges. Do not put your fingers on the curved refracting surfaces. If it is necessary to set the lens down, place it on a clean sheet of paper and never on a tabletop or other surface which may be dusty or greasy.

In order to determine the focal length of the small converging lens, place it in a lens holder and mount it on the optical bench. Place the ground glass screen in another and find the image of a reasonably distant (100 ft or more) object on the screen. In using such a screen, always view the image on the ground surface of the glass. The distance between the lens and the position of the image is nearly equal ot the focal length of the lens in this case. Measure this distance as accurately as possible.

A second method for determining the focal length of a converging lens is called *autocollimation.* Mount the illuminated object and its holder on the optical bench. Also mount the small lens with holder and the plane mirror on the optical bench so that the mirror is located on the opposite side of the lens from the object. Move the lens and the mirror *together* along the bench until a clearly focused, inverted image is formed in the object plane. The image can easily be located with a small white card, but to prevent overlapping of the object and the image the mirror should be tilted very slightly. The distance between the lens and the position of the object-image will be equal to the focal length of the lens. Measure and record this distance.

ANALYSIS

Draw a ray diagram on the report sheet to illustrate the distant-object method of determining the focal length of a converging lens. Explain the method of autocollimation by drawing a ray diagram on the report sheet. Assume that the object coincides with the image at the focal point. (*Hint*: Recall how the focal plane is defined.)

Part II   The Lens Formula

PROCEDURE

The geometrical constructions using the principal rays can be replaced with algebra. The relationlship that exists between the image distance s', object distance s, and focal length f of a given lens is called the *lens formula.* We will examine this relationship experimentally. Use the illuminated arrow on the lamp as the object and use the ground glass screen to locate the real image produced by the smal (short focal length) lens. Measure and record the image distance s and the image height h' for at least ten values of the object distance s. Measure the object height h.

ANALYSIS

Compare the magnification $m = h'/h$ to the ratio s'/s of the image to object distance. Plot s' against s. This is a plot of the lens formula, but it is not easily interpreted. To obtain a simpler graph, plot $1/s'$ against $1/s$ and draw a straight line through the points. Find the two intercepts of this line. From the definition of focal length, the intercepts so obtained are equal to $1/f$. Compute f from the intercepts and compare with the values obtained by the two previous methods. The linear relation assumed represents the lens formula

$$\frac{1}{s} + \frac{1}{s'} = \frac{1}{f}$$

Part III The Refracting Telescope

PROCEDURE

Set up the large (long focal length) lens on the optical bench. Arrange the apparatus so that the image of a distant object can be seen. Observe the real image on the ground glass screen and note its location. Take the small (short focal length) lens whose focal length f has been measured and place it a distance f beyond the image of the large lens. Observe the resulting image by eye.

ANALYSIS

Draw a ray diagram on the report sheet, assuming that the object is at infinity.

Name_____

Date_____

REPORT SHEET
EXPERIMENT 16   LENSES

Part 1   Focal Length of the Converging Lens

DATA AND ANALYSIS

Draw a ray diagram of the distant-image method for determining the focal length f.

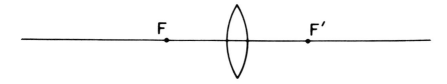

Focal length, $f =$ _____

Draw a ray diagram of the autocollimation method for determining the focal length f.

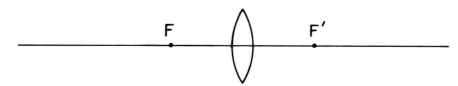

Focal length, $f =$ _____

QUESTIONS

1 In the distant-image method, is the image (*a*) real or virtual and (*b*) inverted or upright? Explain briefly how this method gives the focal length.

2 Does the autocollimation method depend on where the mirror is located?

3 Which of the two values for the focal length f obtained in this **part** of the experiment do you think is more accurate?

Part II   The Lens Formula

DATA

Object height, $h =$ _____

Object distance s (cm)	Image distance s' (cm)	Image height h' (cm)	Magnification $m = s'/s$	h'/h	$1/s$ (cm^{-1})	$1/s'$ (cm^{-1})

ANALYSIS

Complete the above table. Plot s' against s on the graph below and draw a smooth curve through the points.

Plot $1/s'$ against $1/s$ on the graph below and draw a straight line through the points. Find the x and y intercepts of the line.

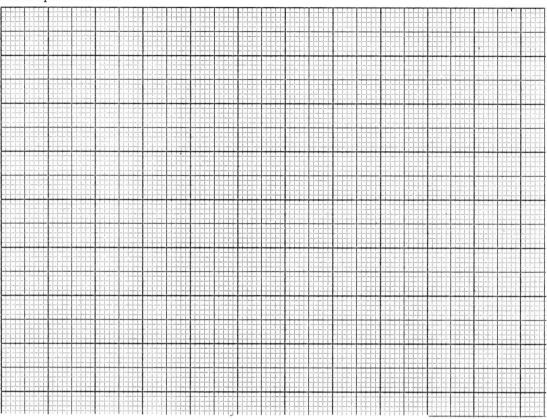

x intercept, $X =$ _____

y intercept, $Y =$ _____

$f = 1/X =$ _____ $= 1/Y =$ _____

QUESTIONS

1. (a) What does s approach as s' increases to infinity?
   (b) What does s' approach as s increases to infinity?
2. Would it make any difference if s and s' were interchanged on your plot?

Part III The Refracting Telescope

DATA AND ANALYSIS

Draw a ray diagram of the refracting telescope. Assume the object is at infinity.

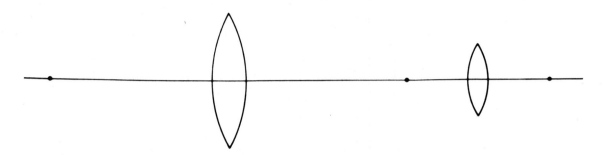

QUESTIONS

1. Is the image in the telescope real or virtual?
2. What role does the eye play here? Draw a ray diagram.

electricity and magnetism

EXPERIMENT 17 THE ELECTRIC FIELD

GOALS

1 To map the electric lines of force between two point charges and two parallel plates.

2 To map the equipotential surfaces between two point charges and two parallel plates.

EQUIPMENT

Two 1' X 1' boards Vacuum-tube voltmeter (VTVM) or equivalent with diode protection
10 V dc power supply Field probe with two contacts mounted 2 cm apart
The two boards are overlaid with graphite-emulsion conducting paper on which electrodes are painted with silver conducting paint and a 1 cm X 1 cm grid is painted with white ink.

INTRODUCTION

A fixed array of electric charges exerts a force on any other charge in its vicinity. For example, a single positive charge $+Q$ exerts a repulsive force on any other positive charge $+q$ (Fig. 17.1). The magnitude of this force, in newtons (N), is given by Coulomb's law

$$F = K\frac{qQ}{r^2} \qquad\qquad 1$$

where Q and q are the magnitudes of the two interacting charges in Coulombs (C), r is the distance between them in meters (m), and $K = 9 \times 10^9$ N-m²/C² is the electric constant. The capacity that the charge has of producing an electric force on another charge in its vicinity can be conveniently described by introducing the concept of a *field*. At any point in space, the electric field **E** of an array of charges is defined as the force that the array would exert on a unit positive charge at that point.

The magnitude of the electric field due to the point charge $+Q$ is $E = F/q = KQ/r^2$. The direction of the field at any point P due to a given array of charges is the direction of the force that would be exerted on a *positive* test charge at P. We represent this direction with an arrow which, in the case of the isolated positive charge $+Q$, points radially outward from $+Q$.

It is not necessary to restrict our attention to the consideration of point P only. In fact, the convenient property of the field concept is that it enables one to determine what would happen to a charge at *any* point in the region surrounding $+Q$. The reason for this is that **E** does not depend on the magnitude of the other charge, but only on the magnitude of the source charge $+Q$, and the distance of separation r.

The electric field surrounding a positive charge $+Q$ can be represented by a diagram such as that in Fig. 17.2. The direction of the arrows gives the field direction at a given point. A qualitative idea of the magnitude of **E** is given by the relative spacing of the field lines; the higher the density of lines, the stronger or more intense is the field. It is important to remember that the lines of **E** emerge from a positive charge and converge on a negative charge.

When a charge q moves a distance d in an electric field, the *work* done by the electric field is

$$W = F_x d = qE_x d \qquad\qquad 2$$

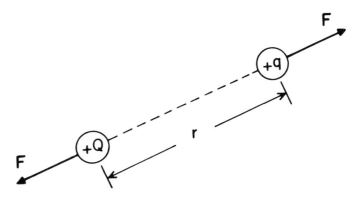

FIGURE 17.1 Two positive charges repel each other.

where E_x is the component of the electric field parallel to d. The *potential difference* $V_{AB} = V_A - V_B$ between two points A and B is equal to the work done by the electric field in moving a *unit charge* from A to B

$$V_{AB} = \frac{W}{q} = E_x d \qquad \qquad 3$$

A surface on which the electric potential is everywhere the same is called an *equipotential*. Since there is no potential difference between any two points on such a surface, the component of **E** parallel to the surface must be zero. This is another way of saying that electric field must be *perpendicular* to the equipotential. In fact, it follows from Eq. 3 that for a fixed distance d, the potential difference is a *maximum* in the direction of **E**.

In this experiment, you will delineate the electric field lines and the equipotential surfaces due to two different arrays of charges. Because static fields are difficult to measure, a feeble electric current (flow of charges) is maintained in a conducting medium between the charge arrays that exist on metal

FIGURE 17.2 Electric field at a point positive charge.

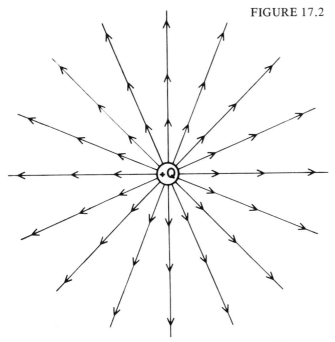

electrodes. The steady-state electric field lines for a given electrode configuration closely resemble the static field that the same configuration of static charges would produce.

REFERENCE Cromer: *Physics for the Life Sciences*, Sec. 16.2, 16.3, and 16.4.

Part I The Electric Dipole

PROCEDURE

The apparatus consists of a flat board which is overlaid with graphite conducting paper imprinted with grid lines. The electrode configuration to be used is the one shown in Fig. 17.3. The voltage between the two contacts on the field probe is measured with a vacuum-tube voltmeter (VTVM) or other high-impedance voltmeter.

Connect the power supply to the two electrode contacts. Set the voltage to 10 V. Set the VTVM to dc volts and connect the field probe to the input of the VTVM. Starting from the 10-V scale, increase the sensitivity of the VTVM until a midscale reading is obtained when the field probe is placed on the paper and oriented along the line joining the electrodes with the (−) contact nearer the negative electrode. Always check the zero of the VTVM after changing scales by shorting the input leads. Make the appropriate adjustment if necessary.

Position the minus (−) contact of the field probe anywhere near the *negative* electrode and, using it as a pivot, slowly rotate the other (+) contact on the paper while observing the VTVM reading. Find the direction of the probe that gives the maximum meter-reading and record the positions of the two probe contacts on the grid included in the report sheet.

Use the second point as the pivot and repeat the above procedure, recording the new, third point on the grid. Continue in this manner until the positive electrode is reached. These points lie along an electric field line. Repeat the entire procedure for two more field lines.

Pick a point somewhere in the center of the paper and place the (−) field-probe contact there. Use this contact as pivot and rotate the other (+) contact to find the position that gives a zero reading on the meter. Record the two points on the same grid sheet, but use a symbol different from the one used to record the electric field. Use the second point as pivot and scan for the next point along the equipotential line. Continue this procedure until an entire equipotential is traced out.

ANALYSIS

Draw a smooth curve through each of the electric-field lines obtained above. Label these lines with arrows that show the direction of the electric field.

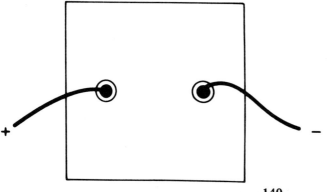

FIGURE 17.3 Electric dipole.

Draw a curve through the equipotential line and confirm that it intersects the field lines at an angle of 90°. Using this fact, sketch in a few more equipotentials.

Part II   Parallel Plates

PROCEDURE

The board to be used in this part has two parallel linear electrodes separated by a distance of 15 cm, as shown in Fig. 17.4. This arrangement can be viewed as a two-dimensional cross-section of a parallel-plate capacitor.

Connect the 10 V supply to the electrode terminals. At a position midway along the length of the electrodes, place the minus (−) contact of the field probe close to the negative plate and scan for the direction giving maximum deflection on the VTVM. Mark the two points on the grid in the report sheet and record the meter reading. Use the second point as pivot and scan for a third point that gives maximum deflection. Again, record the meter reading. Continue this process until the positive plate is reached.

Repeat this procedure for a starting point near an inside edge of one of the plates, and for a starting point near an outside edge of the plates.

At a point just beyond the edge of the plates, initiate the mapping of an equipotential line by using the zero-deflection method.

ANALYSIS

Connect the electric-field points for the three field lines you have mapped. Compare the meter readings obtained in each step and use Eq. 3 to find the value of the electric field in the region between the plates.

Connect the points defining the equipotential line and sketch several other equipotentials.

FIGURE 17.4   Parallel-plate capacitor.

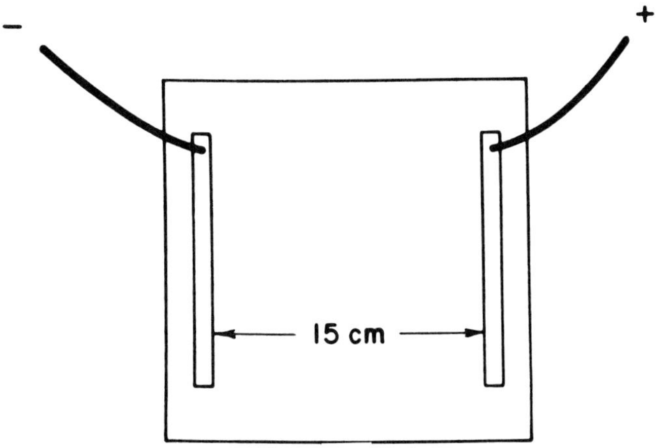

Name_____

Date_____

REPORT SHEET
EXPERIMENT 17   THE ELECTRIC FIELD

Part I   The Electric Dipole

DATA AND ANALYSIS

Plot the points found using the probe, using different symbols for the electric-field lines and the equipotential line. Draw smooth curves through each set of electric-field points. Draw a curve through the equipotential points and confirm that it intersects the field lines at an angle of 90°. Sketch in more equipotentials.

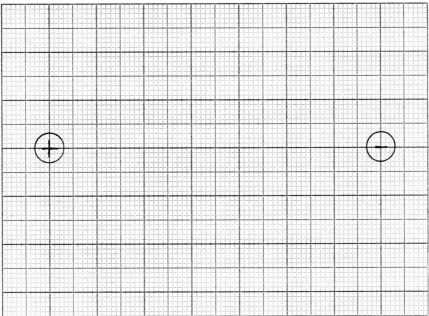

QUESTIONS

1. What is the direction of the force on a positive test charge located 3 spaces above the midpoint of the line connecting the two charges of the dipole?

2. In which region(s) of the plane is the electric force on a test charge the largest? Where is the force the smallest?

Part II Parallel Plates

DATA AND ANALYSIS

Map the three field lines. Record the voltage readings on the grid near the recorded points. Find the value of the electric field in the region between plates. Map one equipotential and sketch several others.

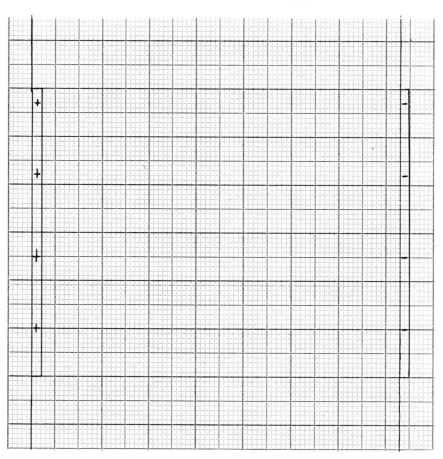

QUESTIONS

1. Does the electric field strength change with position in the region between the plates?
2. At a position midway between the plates a positive charge enters the plates with a velocity that is directed parallel to the plates. Describe its trajectory.

EXPERIMENT 18   OHM'S LAW

GOALS

1 To learn to use an ammeter and a voltmeter.

2 To measure the voltage-current characteristics of ohmic and nonohmic circuit elements.

3 To measure the resistances of several resistors.

4 To determine the characteristics of a series circuit from the characteristics of the individual elements in the circuit.

EQUIPMENT

Current-limited power supply
Ammeter
Voltmeter
Wire leads
Resistors
Small light bulb

INTRODUCTION

An electric current is a flow of charge, and the magnitude of the current I in a wire is the amount of charge that flows through the wire in unit time, or the rate of charge flow. Current is measured in coulombs per second (C/s), or amperes (A)

$$1\,A = 1\,\frac{C}{s}$$

In a metallic conductor, such as a wire, the negatively charged electrons are the only particles that are free to move. To avoid unnecessary minus signs in our equations, the direction of the current in a wire is taken by convention to be the direction in which positive charge would flow. Thus, the conventional current is directed opposite to the direction in which the electrons move.

The flow of charge through a wire is analogous to the flow of a fluid through a capillary (Experiment 8). Just as the fluid flow Q in a capillary depends on the pressure difference Δp between the ends of the capillary, the charge flow (current) I in a wire depends on the potential difference V between the ends of the wire. In the simplest case, I is proportional to V, so we can write

$$I = \frac{V}{R} \qquad\qquad 1$$

where R is the *resistance*, a constant characteristic of the wire. This relation is called *Ohm's law*, and it is the electric analog of Poiseuille's law for fluid flow. Although Ohm's law is valid only for some substances, its simplicity makes it a useful relation with which to begin a study of electric circuits.

REMARK From Eq. 1 we see that $R = V/I$, so the unit of resistance is volts per ampere (V/A). This unit is called the *ohm* Ω

$$1\,\Omega = 1\,\frac{V}{A}$$

Electrical wires have negligible resistance, and so the major resistance of a circuit is due to the specific electrical elements, such as motors, light bulbs, and resistors, that are in the circuit. A *resistor* is an element that obeys Ohm's law to good approximation. In this experiment you will study the properties of simple elements, such as resistors and light bulbs, by simultaneously measuring the current in them and the potential difference across them.

To obtain a constant potential difference (voltage) across a resistor, a device called a seat of emf is used. This device can be either a battery, which is a chemical device that converts chemical energy into electrical energy, or a power supply, which is a electronic device that converts the alternating 120-V line potential into a nonalternating (dc) low voltage. A power supply is more convenient than a battery in electrical experiments, because it provides a very stable voltage that can be varied by turning a dial.

Figure 18.1 shows a resistor R, represented by the symbol ─/\/\/\─, connected to a power supply. The current is measured by an *ammeter* A, which is connected in series with R so that the current in A is the same as in R. (An ammeter itself has negligible resistance.) The voltage across the resistor is measured by a *voltmeter* V which is connected in parallel with R so that the voltage across V is the same as across R. (A voltmeter has a very large resistance so that the current in it is negligible.)

The range of a meter is the current (or voltage) that produces a full deflection of the meter needle. Some meters have a control for selecting different range settings. The face of such a variable-range meter usually has several scales, as shown in Fig. 18.2. The value of the current indicated by the needle in Fig. 18.2 depends on the range setting. If the range setting is 3 mA, the current is found to be 2 mA by reading the position of the needle on the bottom scale. If the range setting is 30 mA, the bottom scale is still used, and the reading is multiplied by 10. If the range control is set at 1 mA, the current is found to be 0.65 mA by dividing the reading on the upper scale by 10. With a little practice you will be able to read these meters without difficulty.

REFERENCE   Cromer: *Physics for the Life Sciences*, Sec. 17.1.

        Part I   **Voltage-current Relations of Single Elements**

PROCEDURE

Connect resistor R_1 to the power supply and meters, as shown in Fig. 18.1. Since an incorrect con-

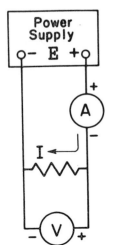

FIGURE 18.1   Connections for measuring the voltage across and the current in a resistor.

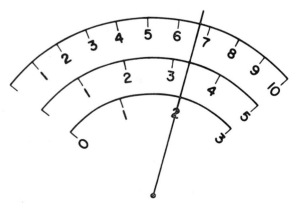

FIGURE 18.2   Scales on a multirange meter.

nection may damage the ammeter, double-check that your connections correspond to Fig. 18.1. Reverse the connections to a meter if its needle deflects to the left when the power supply is turned on.

Measure the current I for ten values of the voltage V between 0 and 10 V. For each measurement, change the range setting on each meter to give the largest possible deflection of the needle.

Repeat the above measurements for resistor R_2, for a light bulb, and for any other element provided by your instructor.

ANALYSIS

Plot the voltage V against the current I for each element. These plots can be made on the same graph, but you should use different symbols (e.g., 0, X, Δ) to distinguish the points for different elements. From Ohm's law the voltage-current relation for a resistor is

$$V = R I$$

so that the points for each resistor should fall on a straight line that passes through the origin. Determine the resistance of the resistors from the slopes of the straight lines drawn through the points.

The points for the light bulb do not fall on a straight line because the resistance of the light bulb increases as it gets hotter. Draw a smooth curve through these points. The V–I characteristics of a resistor are specified by a single number, its resistance R, but to specify the V–I characteristics of a non-ohmic element, such as a light bulb, the entire V–I curve of the element must be given. From such a V–I curve, it is possible to determine how the element will operate in any circuit.

Part II   Voltage-current Relation of Two Elements in Series

PROCEDURE

Connect the resistor R_1 in series with the light bulb. Measure the current I in this composite element for various values of the voltage V across it.

ANALYSIS

Plot V against I, and connect the points with a smooth curve.

The V–I, curve for two elements in series can also be calculated from the V–I curves of the individual elements. For a given value of I, the separate voltages V_1 and V_2 across R_1 and the light bulb, respectively, are found from the V–I curves in Part I. When these elements are connected in series, the current I is the same in both of them, so the potential across the two together is $V = V_1 + V_2$. Calculate $V = V_1 + V_2$ from the V–I curves in Part I for five values of I. Plot these V–I values on the graph with the measured V–I curve for the two elements in series. The calculated points should lie on the measured curve.

Name_____

Date_____

REPORT SHEET
EXPERIMENT 18   OHM'S LAW

Part I   Voltage-Current Relations of Single Elements

DATA

R_1		R_2		Light bulb			
V (V)	I (mA)	V (V)	I (mA)	V (V)	I (mA)	V (V)	I (mA)

ANALYSIS

Plot V against I for each element.

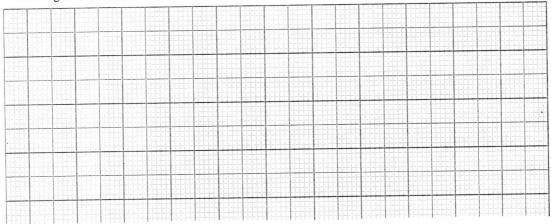

Draw a straight line through the points for each resistor, and determine the resistances from the slopes of the lines.

$R_1 = $ _____

$R_2 = $ _____

149

QUESTIONS

1 The voltage across a resistor is 7.5 V when the current in it is 120 mA. What is the resistance of the resistor?

2 What will be the voltage across R_1 when the current in it is 2.5 A?

3 How can you tell from your data whether the resistance of the light bulb increases or decreases with increasing current?

Part II   Voltage-Current Relation of Two Elements In Series

DATA

R_1 and light bulb in series	
V (V)	I (mA)

ANALYSIS

Plot V against I, and connect points with a smooth curve.

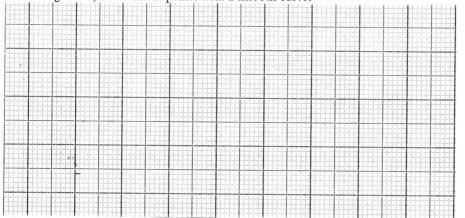

From the data in Part I, calculate $V = V_1 + V_2$ for five values of I. Plot these calculated $V - I$ points on the above graph.

QUESTIONS

1 If R_1 and R_2 were connected in series, what would be the voltage across the pair when the current is 12 mA?

2 A resistor R and the light bulb are connected in series. If the current is 6 mA when the voltages across the two elements is 8 V, what is R?

EXPERIMENT 19 KIRCHOFF'S LAWS

GOAL

To compare the currents and voltages measured in various circuits to the values calculated using Kirchhoff's laws.

EQUIPMENT

Current-limited power supply
Voltmeter
Ammeter
Three resistors
Wire leads

INTRODUCTION

The current in each branch of a complex circuit network can be calculated by a systematic application of the following two principles, known as *Kirchhoff's laws*:

1 The total current entering any point of a circuit is equal to the total current leaving the point.

2 The potential difference between any two points in a circuit is the same along any path connecting the points.

We first apply these principles to the simple series circuit in Fig. 19.1. According to the first law, the current I is the same in all the resistors, so the voltage across the individual resistors are

$$V_1 = IR_1$$
$$V_2 = IR_2 \qquad\qquad 1$$
$$V_3 = IR_3$$

and the total potential difference between points a and b (going through the resistors) is

$$V = V_1 + V_2 + V_3 = I(R_1 + R_2 + R_3)$$

Going through the seat of emf, the potential difference between a and b is $\mathcal{E}$. From Kirchhoff's second law we have $V = \mathcal{E}$, so

$$I = \frac{\mathcal{E}}{R_1 + R_2 + R_3} \qquad\qquad 2$$

In general, n resistors in series act as a single resistor of magnitude

$$R = R_1 + R_2 + \cdots + R_n$$

From Eqs. 1 and 2, the current in the circuit and the potentials across the resistors can be calculated in terms of $\mathcal{E}, R_1, R_2,$ and R_3.

In the parallel circuit in Fig. 19.2, the total current I splits into three parallel currents. From Kirchhoff's first law we have

$$I = I_1 + I_2 + I_3$$

From the second law, the emf $\mathcal{E}$ of the power supply is equal to the potential across each resistor, so

$$I_1 = \frac{\mathcal{E}}{R_1}$$
$$I_2 = \frac{\mathcal{E}}{R_2} \qquad \qquad 3$$
$$I_3 = \frac{\mathcal{E}}{R_3}$$

Adding these last three equations we get

$$I = I_1 + I_2 + I_3 = \mathcal{E}\left[\frac{1}{R_1} + \frac{1}{R_2} + \frac{1}{R_3}\right] \qquad 4$$

In general, n resistors in parallel act as a single resistor whose magnitude R is given by

$$\frac{1}{R} = \frac{1}{R_1} + \frac{1}{R_2} + \cdots + \frac{1}{R_n}$$

From Eqs. 3 and 4, the currents I, I_1, I_2, and I_3 can be calculated in terms of $\mathcal{E}, R_1, R_2$, and R_3.

By a similar application of Kirchhoff's laws, you should be able to show that the currents in the branches of the circuit in Fig. 19.3 are

$$I_1 = \frac{\mathcal{E}}{R_1}$$
$$I_2 = \frac{\mathcal{E}}{(R_2 + R_3)} \qquad \qquad 5$$

and that the total current is

$$I = I_1 + I_2 = \mathcal{E}[\frac{1}{R_1} + \frac{1}{(R_2 + R_3)}] \qquad 6$$

The circuit in Fig. 19.4 is most easily analyzed by first calculating the total resistance of the circuit. The two parallel resistors act as a single resistance of magnitude

$$\frac{1}{R'} = \frac{1}{R_2} + \frac{1}{R_3} = \frac{R_2 + R_3}{R_2 R_3}$$

or

$$R' = \frac{R_2 R_3}{(R_2 + R_3)}$$

This resistance is in series with R_1, so the total resistance of the circuit is

$$R = R_1 + R' = R_1 + \frac{R_2 R_3}{(R_2 + R_3)}$$

and the current I is

$$I = \frac{\mathcal{E}}{R} = \frac{\mathcal{E}}{[R_1 + (R_2 R_3/R_2 + R_3)]} \qquad 7$$

The potential across R_2 and R_3 is $\mathcal{E} - R_1 I$, so the currents in R_2 and R_3 are

$$I_2 = \frac{\mathcal{E} - R_1 I}{R_2}$$

$$I_3 = \frac{\mathcal{E} - R_2 I}{R_3}$$

8

In this experiment you will build the four circuits shown in Figs. 19.1 through 19.4, measure the currents in and the voltages across each resistor, and compare the measured values with the values calculated from Eqs. 1 through 8.

REFERENCE Cromer: *Physics for the Life Sciences*, Sec. 17.2 and Experiment 18.

PROCEDURE

You will be supplied with three resistors, R_1, R_2, and R_3. With the power supply, voltmeter, and ammeter connected as in Fig. 18.1, measure the current in each resistor for several values of the voltage across the resistor. These data will be used to determine the resistance of each resistor.

1 Connect the three resistors in series with the power supply (Fig. 19.1), and measure the current I in the circuit, the potentials V_1, V_2, and V_3 across R_1, R_2, and R_3, and the emf $\mathcal{E}$ of the power supply.

2 Connect the three resistors in parallel with the power supply (Fig. 19.2), and measure the currents I, I_1, I_2, and I_3, and the emf ϵ.

3 Connect the three resistors to the power supply as shown in Fig. 19.3. Measure the currents I, I_1, and I_2, the potentials V_2 and V_3 across R_2 and R_3, and the emf $\mathcal{E}$.

4 Connect the three resistors to the power supply as shown in Fig. 19.4. Measure the currents I, I_2, and I_3, the potentials V_1 and V_2 across R_1 and R_2, and the emf $\mathcal{E}$.

ANALYSIS

From the V-I data of the individual resistors, determine the resistance of each resistor.

For each circuit in Figs. 19.1 through 19.4, calculate in terms of R_1, R_2, R_3, and $\mathcal{E}$ the voltages and currents that you measured. The calculated values should agree with the measured values within the accuracy of the meters (about 3%).

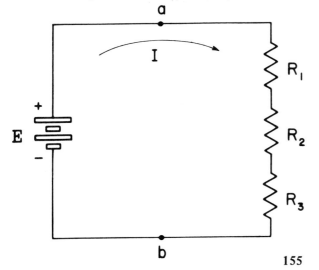

FIGURE 19.1   Three resistors in series.

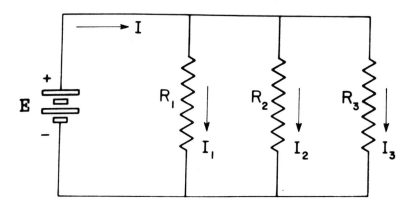

FIGURE 19.2   Three resistors in parallel.

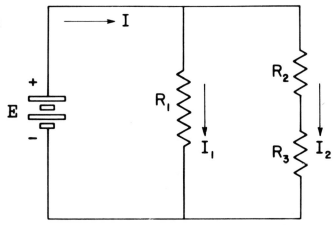

FIGURE 19.3   A resistor in parallel with two resistors in series.

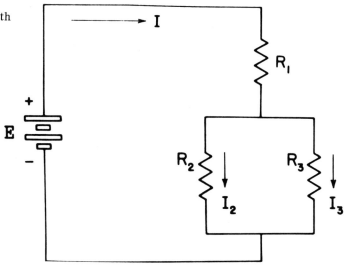

FIGURE 19.4   A resistor in series with two resistors in parallel.

Name_____

Date_____

REPORT SHEET
EXPERIMENT 19   KIRCHHOFF'S LAWS

DATA AND ANALYSIS

Record the current in each resistor for several values of the voltage. Determine the resistance of each resistor from these data.

R_1		R_2		R_3	
V (V)	I (mA)	V (V)	I (mA)	V (V)	I (mA)

$R_1 =$ _____

$R_2 =$ _____

$R_3 =$ _____

Record the measured currents and voltage for each circuit studied. Calculate these same quantities from $\mathscr{E}$ and the above values of the resistances.

Circuit 1		
	measured	calculated
$\mathscr{E}$		
I		
V_1		
V_2		
V_3		

Circuit 2		
	measured	calculated
$\mathscr{E}$		
I		
I_1		
I_2		
I_3		

Circuit 3	measured	calculated
$\mathcal{E}$		
I		
I_1		
I_2		
V_2		
V_3		

Circuit 4	measured	calculated
$\mathcal{E}$		
I		
I_2		
I_3		
V_1		
V_2		

QUESTIONS

1. In each of the four circuits, describe how the current I is related to the currents $I_1, I_2,$ and I_3 in each resistor.

2. In each of the four circuits, describe how the emf is related to the voltages $V_1, V_2,$ and V_3 across each resistor.

3. From your measured values of $\mathcal{E}$ and I, determine the total resistance of each circuit. Compare these to the total resistances calculated from the known values of $R_1, R_2,$ and R_3.

4. Which combination of three resistors gives the smallest total resistance?

EXPERIMENT 20   THE OSCILLOSCOPE

GOALS

1. To learn to make measurements with an oscilloscope.
2. To understand the standard terminology applied to ac circuits and waveforms.
3. To study the relationship between the perception of sound by the human ear and the physical attributes of sound (intensity and frequency).
4. To obtain and study an electrocardiogram obtained from a human subject.

EQUIPMENT

50-kHz-to-dc-coupled oscilloscope
6-V battery
6.3-V ac step-down (filament) transformer
110 V ac mercury-wetted relay
Audio oscillator (10 Hz to 50 kHz)

Audio-output transformer
Hi-fi speaker
High-impedance differential amplifier
ECG electrodes and paste

INTRODUCTION

Among the many different sources of electrical excitation, the simplest are those that supply *direct current* (dc). A direct current source, such as a battery, provides a steady current that passes in a fixed direction through any external circuit connected to it. It is, therefore, only necessary to know the polarity and the magnitude of the source in order to predict from Kirchhoff's laws what will happen in a particular circuit. The instruments used for measuring direct current, such as ammeters and voltmeters, are all based on the D'Arsonval galvanometer, which responds to the unidirectional current in it.

In contrast to the battery there exist sources that produce a current that varies with time both in magnitude and in direction. Most bioelectrical signal sources fall into this category. An important subclass of such sources are those that are *periodic*: that is, they produce a variable current that repeats itself in a fixed interval of time called the *period*. The most common practical periodic sources produce a current that is a sinusoidal function of time. This is called an *alternating current* (ac) because it changes both direction and magnitude periodically as shown in Fig. 20.1.

There are several common terms used in referring to a signal of this type. These are indicated in Fig. 20.1 and their definitions are listed below:

1. The amplitude A is the maximum value of the ac voltage or current measured in volts or amperes, respectively.
2. The peak-to-peak (p-p) amplitude is the voltage or current difference between the top of the "hills" and the bottom of the "valleys". In the case of a symmetrical waveform, it is equal to twice the amplitude A.
3. The period τ is the time taken to complete one full cycle, measured in seconds.
4. The frequency f is the number of cycles that are completed per unit time. The unit of frequency is the Hertz (Hz), which is the number of cycles per second. Frequency is the reciprocal of the period.

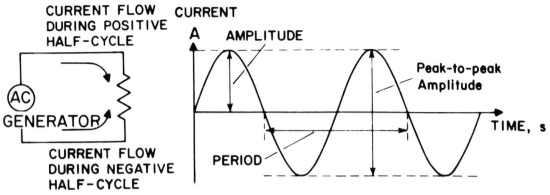

FIGURE 20.1   Alternating current waveform.

Although there are direct-reading meters made especially to respond to alternating current, usually only one of the above quantities, the amplitude, can be determined from such a measurement. In order to measure the period (frequency), the shape, and other characteristics of a periodic signal, it is necessary to use an instrument called the *oscilloscope*.

The oscilloscope actually traces out the time-dependent waveform of the signal being measured. The face of the oscilloscope is the screen of a cathode-ray tube, which is similar to the picture tube in a television set. The input signal is amplified and then used to control the vertical position of a beam of electrons that provides the trace observed. The horizontal position of the beam is proportional to time because it is swept across the screen at a fixed speed. The use of the oscilloscope is not, of course, restricted to the observation of periodic signals; the instrument can also be employed to measure and record quasi-periodic signals, such as exist in biological systems, or even of transient signals, which occur as single or random events. In this experiment an introduction to the operation and capabilities of the oscilloscope is presented with some applications.

REFERENCE Cromer: *Physics for the Life Sciences*, Sec. 16.5, 13.3, and 17.5.

Part I   Operation of the Oscilloscope

PROCEDURE

The apparatus used in this experiment consists of a single-trace 50-kHz oscilloscope (50 mV/cm sensitivity), a 6-V battery, a 6.3-V ac step-down filament transformer (with at least 10-kΩ in series with the secondary and the 5-kΩ load), and a 110-V ac mercury-wetted relay, which is run off the ac line.

The main controls on the oscilloscope are the intensity or brilliance knob, the vertical amplifier scale knob (usually a multi-position switch calibrated in terms of volts per centimeter of vertical deflection on the screen), the sweep-time knob (usually a multi-position switch calibrated in terms of seconds per centimeter of horizontal displacement on the screen), and the coupling switch, which determines whether the input amplifier of the oscilloscope will respond to dc or purely ac signals.

Set the vertical-amplifier-scale knob to the 5-V/cm position. (Be sure the calibration switch is set properly.) Set the sweep-time knob to the 5-ms/cm position. Use dc coupling. Turn the instrument on and adjust the intensity control so that a trace of moderate brightness is obtained. (*Caution*: Always maintain as little intensity as is necessary to give good viewing, because a beam with excessive intensity can permanently damage the phosphor screen). It may be necessary to adjust the focus control to obtain a clear trace. If any difficulties arise at this point, consult your instructor immediately.

With the input shorted or on *ground* position, center the trace on the screen by using the vertical- and horizontal-position controls. Connect the 6-V battery to the input (switch from ground to dc coupling), with the negative terminal attached to the ground lead and the positive pole attached to the signal lead. Observe the result and measure the deflection. Express the deflection in volts. Change the vertical-scale setting so that the most accurate value of the voltage can be obtained (i.e., the lowest V/cm-scale setting that still leaves the trace on the face of the screen). Be sure to recenter the trace every time the vertical-scale setting is changed. Reverse the lead connections to the battery. Observe and record the results.

Connect the 6-V cell to the NC (normally closed) and NO (normally open) terminals of the relay box. Attach the oscilloscope leads to the common and NO terminals, respectively. Plug the relay box into an ac outlet. Observe the waveform and draw a sketch of it on the report sheet. Measure the p-p amplitude and the period of this signal by determining how many centimeters one cycle takes up on the screen. Set both the vertical-scale knob and the sweep-time knob for the best accuracy. Pay special attention to the position of the waveform relative to the zero-volt level. Switch to ac coupling and compare with the trace observed using dc coupling. Record your observations on the report sheet.

Next, connect the secondary output terminals of the filament transformer to the oscilloscope. Switch back to dc coupling. Plug the transformer primary into an ac outlet. (*Caution*: Do not touch any of the terminals without unplugging the transformer first.) The transformer allows you to observe the voltage waveform at the power line, but at a voltage much reduced from the approximately 300-V p-p level present there. Make a sketch of the waveform on the report sheet. Measure and record both the p-p amplitude and the period of the observed signal. Switch to ac coupling and compare with the previous signal.

ANALYSIS

Compute the frequency for the two ac signals observed, and briefly describe the similarities and differences between them. Any periodic signal can be expressed as the sum of a dc signal equal to the average value of the periodic signal and an ac signal with an average value of zero. Find the dc or average value for each waveform observed by drawing the horizontal line that divides the waveform in half with equal areas above and below the line. Explain what is accomplished by switching from dc to ac coupling.

Part II   Psychophysics of Sound

PROCEDURE

The apparatus used in this part of the experiment consists of an audio oscillator with a frequency range of at least 10 Hz to 50 kHz, an audio-output transformer (600 Ω primary to 8 Ω secondary), and an inexpensive styrofoam high-fidelity loudspeaker.

Connect the primary leads of the transformer to the audio-oscillator-output terminals, and the secondary transformer leads to the loudspeaker terminals as illustrated in Fig. 20.2. Also connect the oscilloscope input leads to the speaker terminals. Make sure that the level or output control on the oscillator is turned fully counterclockwise. Set the frequency dial to 300 Hz and turn on the oscillator. Increase the oscillator level control until a moderately loud sound is produced by the speaker. Measure the p-p amplitude of the voltage impressed on the speaker with the oscilloscope. Decrease the oscillator level until the amplitude is reduced by a factor of ten. Record your qualitative impression of how much weaker the sound is. Again decrease the voltage level by a factor of ten and record your perceptual impression. Continue to decrease the voltage by factors of ten until you can no longer hear it. Record the level at which this occurs and compare with the original signal level.

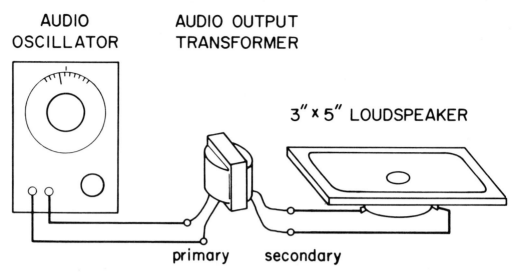

FIGURE 20.1   Curcuit for studying sound.

Restore the oscillator level to a moderate value. Tune the frequency dial toward the high-frequency range, continually maintaining a reasonably constant voltage across the speaker by means of the oscillator output control. Determine the highest frequency at which you can still hear sound. Repeat the procedure to find the lowest frequency at which you can still hear sound. (Due to a decrease in the low-frequency response of the speaker, it will be necessary to increase the oscillator output somewhat.) Record these values.

ANALYSIS

Describe and discuss the range of sensitivity of the human ear with regard to sound intensity and frequency. Bear in mind that for each decrease in the driving voltage by a factor of ten, the intensity of the sound waves is reduced by a factor of approximately 100.

Part III   Electrocardiography

The oscilloscope provides a useful means for observing the feeble bioelectrical potentials that accompany the extension and contraction of animal muscles. One muscle that produces a regular, quasi-periodic pattern is the heart. The process of measuring and graphically recording the heart's electrical activity is termed *electrocardiography*, and the record obtained is called an *electrocardiogram* (ECG or EKG). In many monitoring situations, however, a permanent record of the waveform is not needed, so an oscilloscope is used to display it. Such an instrument is called an ECG monitor and when used in a hospital it is normally coupled to an alarm system to warn attendants of any abnormal changes in a patient's heart rhythm.

PROCEDURE

The apparatus to be used in this experiment consists of the oscilloscope, a fairly high-impedance differential amplifier[†] capable of a gain of 100 or better, and a set of ECG electrodes with paste, or a suitable substitute.

[†] The circuit diagram of a differential amplifier, which can easily be constructed using inexpensive integrated circuit components, is given in the Instructor's Guide.

Connect the electrodes to the right arm (RA), right leg (RL), and left arm (LA) of the subject. Connect the RA and LA leads to the input of the differential amplifier, and the RL lead to ground. The oscilloscope is connected to the output of the differential amplifier. Use ac coupling and set the sweep time to 0.5 s/cm to observe the ECG waveform. Make a sketch of the waveform on the report sheet. Measure the period of the ECG signal and determine its amplitude.

Have the subject make a fist or flex his arm muscles and observe the effects on the oscilloscope. Record these observations.

ANALYSIS

Determine the heartbeat rate (the frequency) from the data. If the gain of the differential amplifier is known, compute the actual voltage present at the ECG electrodes from the measured amplitude on the oscilloscope.

REMARK Be sure to take all electrical safety precautions, including good grounding of the oscilloscope, before attaching electrodes to a human subject.

Name _____

Date _____

REPORT SHEET
EXPERIMENT 20   THE OSCILLOSCOPE

Part I   Operation of the Oscilloscope

DATA AND ANALYSIS

Measurements on 6-V battery

Deflection _____ cm = _____ V

Leads reversed

Deflection _____ cm = _____ V

Sketch the trace below.

Average level = _____ V

Measurements on the battery with relay box

Sketch the waveform below (dc coupling).

p-p amplitude = _____ V   Average, or dc value = _____ V

Period = _____ s   Frequency = _____ Hz

Sketch the waveform obtained with ac coupling below.

Average or dc value = _____ V

Measurements on the filament transformer

Sketch the waveform below.

p-p amplitude = _____ V   Average, or dc value = _____ V
Period = _____ s   Frequency = _____ Hz

QUESTIONS

1 What is accomplished by switching from dc to ac coupling?

2 Which of the three signals measured above was pure dc and which was pure ac?

3 What is the highest frequency you can measure with this oscilloscope? What is the lowest frequency?

Part II   Psychophysics of Sound

DATA AND ANALYSIS

Oscillator level (speaker voltage, V)	Loudness (subjective impression)

Low-frequency limit _____ Hz   High-frequency limit _____ Hz

Discussion:

QUESTIONS

1 Which physical aspect of the sound waves produced by the speaker is associated with loudness and which is associated with pitch?

2 Over how many orders of magnitude in sound intensity is your ear sensitive?

Part III   Electrocardiography

DATA AND ANALYSIS

Sketch the observed waveform below.

Effects observed when muscles are flexed:

Heartbeat period = _____ s   Heartbeat rate = _____ Hz
Observed amplitude = _____ V ÷ amplifier gain = ECG potential = _____ V

QUESTIONS

1 Can the electrical signal produced by flexing the arm muscles be characterized as periodic? Explain.
2 If the RA and LA electrodes attached to a subject are connected, is his heart "shorted out"? Explain.

EXPERIMENT 21   THE MAGNETIC FIELD

GOALS

1 To observe the magnetic fields produced by different magnets.

2 To study the effect of a magnetic field on a bar magnet.

3 To measure the magnitude of the horizontal component of the earth's magnetic field.

EQUIPMENT

2' × 2' transparent plastic board
Iron filings
Bar magnet
C-magnet
10-turn solenoid
Overhead projector
Jar magnet
Helmholtz coils

Cardboard template for Helmholtz coils
Shimming blocks
dc-power supply (0–5 A)
Ammeter (0–5 A)
Reversing switch
22-ohm (5 A) rheostat
Timer
Compass

INTRODUCTION

The opposite ends or *poles* of a magnet are designated north (N) and south (S), just as the two kinds of electric charge are designated positive (+) and negative (−). As with charge, opposite poles attract each other and like poles repel each other. Furthermore, the effect that a magnet or group of magnets has on a pole is conveniently described in terms of the magnetic field **B**, which is analogous to the electric field **E**.

The magnetic field at a point P due to a magnet is equal to the force exerted by the magnet on a unit north magnetic pole located at P. Thus, the magnetic field is a vector quantity that can be associated with every point in the vicinity of the magnet. It can be graphically represented by drawing the *field lines* associated with the magnet (Fig. 21.1). At any point P the direction of **B** is tangent to the field line at P and the magnitude of **B** is indicated by the spacing of the lines: the closer the lines are spaced, the greater the magnitude of **B**.

In the case of a C-magnet the direction of **B** is from the north to the south pole and the lines are more or less equally spaced, which indicates that the magnitude of **B** is constant throughout the region between the poles. In the case of a bar magnet, the direction and magnitude of **B** vary throughout the surrounding space in a complicated way. Obviously the magnitude of **B** is greatest near the poles of the magnet.

The field lines produced by a particular source can be determined by using a small bar magnet (such as a compass needle) mounted on a pivot so that it is free to rotate. The magnet responds to a magnetic field at its position by lining up with the field lines. The reason for this is that an external magnetic field produces a torque τ on the bar magnet that is proportional both to the field strength and to the sine of the angle θ between the bar magnet and the direction of **B**

$$\tau \propto B \sin \theta \qquad\qquad 1$$

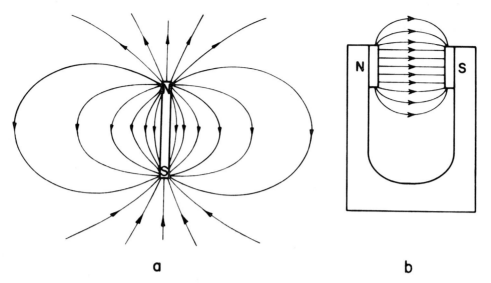

FIGURE 21.1   Magnetic fields of (*a*) a bar magnet and (*b*) a C-magnet.

To be more precise, let us draw an arrow from the south to the north pole of the bar magnet. This arrow represents the *magnetization vector* **M** of the bar magnet. As shown in Fig. 21.2 the magnetic torque on the bar magnet always acts to align **M** with **B**. This kind of torque is called a *restoring torque* because it tends to restore the magnet to the aligned position.

If the bar magnet is already lined up with **B**, the torque is zero and the magnet will remain in that orientation. If the bar magnet is not initially lined up with **B**, however, the restoring torque will cause the bar magnet to oscillate about the direction of **B**. The *period of oscillation T* of the magnet can be shown to be inversely proportional to the square root of the field strength

$$T \propto \frac{1}{\sqrt{B}} \qquad\qquad 2$$

Therefore, the stronger the field is, the shorter the period is (i.e., the faster the magnet oscillates). These oscillations eventually die out because of friction.

This is essentially the way a compass works. The compass needle is simply a bar magnet and its north pole is usually designated by a colored mark. The north pole of the needle always points north, because the earth itself produces a magnetic field to which the compass needle responds. In fact, the earth is likened to a giant bar magnet, which is approximately aligned with the rotational axis. It is apparent that the way we distinguish between the poles of a bar magnet is to see to which magnetic pole of the

FIGURE 21.2   Torque on a bar magnet in a uniform magnetic field **B**.

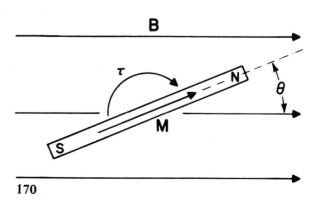

earth they point. Thus, the colored end of a compass needle and the end of a bar magnet marked "N" are properly called the north-seeking poles, because they point to the north magnetic pole of the earth.

REFERENCE Cromer: *Physics for the Life Sciences*, Sec. 18.1 and 18.2.

DEMONSTRATION

The equipment consists of a 2' × 2' clear plastic board, a supply of iron filings, a bar magnet, a C-magnet, a 10-turn solenoid (# 10 AWG copper wire suggested) mounted in such a way that the plastic board cuts it in longitudinal section, a dc power supply (5 A), and an overhead projector.

Your instructor will demonstrate the fields of the bar magnet, the C-magnet with flat pole faces, and the electromagnet (solenoid), by using iron filings, which behave like tiny bar magnets when placed in a magnetic field. On the report sheet draw a sketch of the field lines produced by each one of these objects. Of course, the iron filings do not actually trace out the field lines, but they do respond to the field at their location by lining up with it. The field lines themselves are simply a way of *representing* the physical situation.

The electromagnet mentioned in this demonstration does not contain iron or any other magnetic material, only a coil of wire with electric current in it. Nevertheless, the iron filings indicate that a magnetic field is produced. This interesting connection between moving electrons (i.e., electric current) and magnetism is of fundamental significance. In this experiment, you will use a current-generated magnetic field to calibrate a simple bar magnet device, and then use the device to measure the magnitude of the earth's magnetic field.

> Part I   Calibration of the Jar Magnet

PROCEDURE

The apparatus includes a pair of Helmholtz coils, a cardboard template with cutouts for the coils, shimming blocks, iron filings, a small compass, a jar magnet (a small bar magnet about 5 cm long × 5 mm diameter with one end painted, which is suspended from the lid of a glass baby food jar by a nylon thread), a 5 A dc power supply, a timer, an ammeter (0 to 5 A), reversing switch, and a 22-ohm (5 A) rheostat. Connect the circuit as shown in Fig. 21.3. With the output knob of the supply turned fully counterclockwise (CCW), turn on the supply and increase the output until the ammeter reads 0.5 A. Place the cardboard mask over the Helmholtz coils and arrange the shimming blocks so that the board is located in the horizontal plane that bisects the coil cross section. Sprinkle some iron filings sparingly all over the cardboard sheet. Tap the sheet gently several times so that the filings can line up with the field. Take the compass and determine the field direction at several points inside and outside the coils. Recall that the north-seeking end points in the direction of **B**. Make a sketch of some of the field lines produced by the Helmholtz coils on the report sheet.

Reverse the current by flipping the reversing switch and repeat the above procedure. Sketch the field lines in this case and note any differences. Turn the power supply off, remove the cardboard template and replace the iron filings in their container.

Place the jar magnet in the center of the Helmholtz pair at a horizontal level that is close to the longitudinal axis of the coils. Determine magnetic north with the jar magnet and rotate the Hemholtz coils so that their axis points in that direction.

Turn on the power supply and set the current at 0.5 A. Measure and record the time for the jar magnet to perform 50 oscillations. The period of oscillation is equal to the elapsed time in seconds divided by 50. Repeat this measurement for coil currents of 1.0, 1.5, 2.0, and 2.5 A. When you have completed

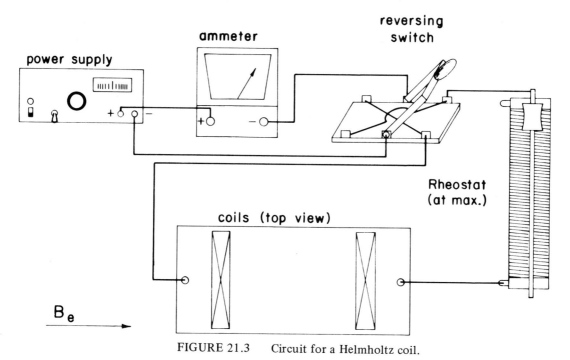

FIGURE 21.3 Circuit for a Helmholtz coil.

the data, turn the power supply off.

ANALYSIS

From Eq. 2 we see that the magnitude of the total magnetic field present at the position of the jar magnet is given by

$$B_{total} = \frac{c}{T^2} \qquad 3$$

where c is a constant of proportionality, which when known serves to provide a *calibration* of the magnitude of a magnetic field in terms of the period of the jar magnet. The magnetic field in teslas (T) produced by the Helmholtz coils is

$$B_H = (9.0 \times 10^{-7} \frac{\text{T-m}}{\text{A}}) \frac{nI}{r} = kI \qquad 4$$

where n is the number of turns per coil, r is the radius (in meters) of each coil, and I is the current (in amperes) in the coils.

Combining Eqs. 3 and 4, we find

$$\frac{1}{T^2} = (\frac{1}{c})(B_H + B_e) = (\frac{1}{c})(B_e + kI) \qquad 5$$

where B_e is the magnetic field of the earth.

Plot $1/T^2$ against I. Draw the best straight line through the points and find the slope s. Compute the value of k from the given physical characteristics of the Helmholtz coils. Using this value and the value obtained for the slope, compute

$$c = \frac{k}{s} \qquad 6$$

The jar magnet is now calibrated so that the magnitude of a magnetic field in which it is placed can be determined directly from the period of oscillation by using Eq. 3.

Part II The Earth's Magnetic Field

PROCEDURE

Set the jar magnet as far as practicable from any steel or iron items (especially the steel girders that may be in the walls). Measure and record the time for 100 oscillations.

ANALYSIS

Compute the period and use Eq. 3 to compute the magnitude of the horizontal component of the earth's magnetic field. Compare the measured value with the value listed for your location in a handbook of physical measurements.

Name_____

Date_____

REPORT SHEET
EXPERIMENT 21   THE MAGNETIC FIELD

DEMONSTRATION

Sketch the field lines observed using the bar magnet, the C-magnet, and the electromagnet.

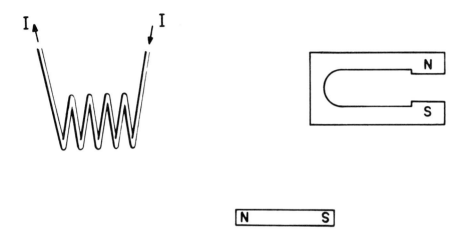

QUESTIONS

1. Assuming that the earth's magnetic field can be represented by a bar magnet, which letter, N or S, designates the end of the magnet nearest the earth's North geographic pole?

2. Explain the torque exerted by a magnetic field on a bar magnet in terms of the attraction and repulsion of magnetic poles.

Part I   Calibration of the Jar Magnet

DATA AND ANALYSIS

Sketch the magnetic field lines of the Helmholtz coils.

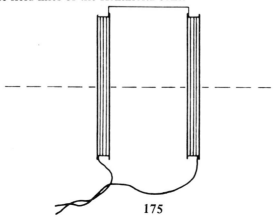

Describe the effects observed when the current in the coils was reversed.

Record the period T of the jar magnet for various values of the current I.

Current I (A)	Time for 50 oscillations (s)	Period T (s)	T^2 (s^2)	$1/T$ (s^{-1})

Physical characteristics of the Helmholtz coils:

    Number of turns per coil, $n =$ _____ turns

    Radius, $r =$ _____ m

Plot $1/T^2$ against I on the graph below:

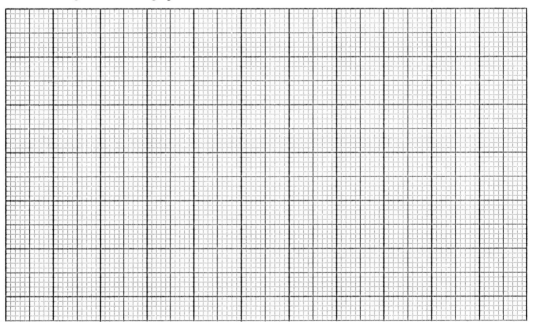

Draw the best straight line through the data points and find the slope, s:

$s =$ _____ $\dfrac{1}{\text{A-s}}$

Compute the field constant of the coils, k, where

$$k = (9 \times 10^{-7}\,\tfrac{\text{T-m}}{\text{A}})(\tfrac{n}{r}) = \underline{\hspace{2cm}} \tfrac{\text{T}}{\text{A}}$$

Compute the calibration constant c:

$c = \dfrac{k}{s} =$ _____ T-s^2

QUESTIONS

1 Explain why the plot of $1/T^2$ against I does not pass through the origin.

2 What is the limitation of the jar magnet for measuring very intense magnetic fields?

Part II   The Earth's Magnetic Field

DATA

Elapsed time for 100 oscillations = _____ s

Period, T = _____ s

ANALYSIS

Calculate magnitude of the earth's field.

$B_e = \dfrac{c}{T^2} =$ _____ T

QUESTIONS

1 How does the value for B_e obtained above compare with the value quoted in the handbook?

2 You have measured only the horizontal component of the earth's magnetic field. There is also a rather large component in the vertical (downward) direction. Why doesn't the jar magnet respond to this field component?

modern physics

EXPERIMENT 22 ATOMIC SPECTRA

GOALS

1 To calibrate a spectroscope.

2 To measure the visible spectral lines of hydrogen and compare them to the predictions of the Bohr model of the atom.

3 To observe the Fraunhofer lines of the sun.

EQUIPMENT

Prism spectrometer Helium and hydrogen spectrum tubes with power supply
Sodium vapor lamp Glass filters, red and blue
Incandescent lamp Colored pencils

INTRODUCTION

When an electric current passes through a gas, the atoms of the gas emit light at specific discrete wavelengths. The spectrum of this light is characteristic of the atoms involved, and gives important information about the arrangement of the electrons in the atom. The hydrogen spectrum, for example, is different from that of helium or any other atom. The Bohr model of the atom pictures an atom pictures an atom as a positively-charged nucleus with a swarm of negatively-charged electrons orbiting around it. However, it was possible for Bohr to explain the discrete nature of atomic spectra only after he imposed a quantum condition on the allowed orbits. This condition restricts the electrons in an atom to specific energy states. The transition of an electron from one state to another involves a definite change in the energy of the atom.

Bohr showed that the energy states of the hydrogen atom are given by the simple formula

$$E_n = \frac{E_1}{n^2} = -\frac{13.6 \text{ eV}}{n^2} \qquad 1$$

where n is an integer ($n = 1, 2, 3, \ldots$) called the *principal quantum number*. The lowest-energy state (called the *ground state*) has an energy $E_1 = -13.6$ eV. When an electron in the ground state is excited by electrical discharge in the gas, it is raised to one of the higher energy states with $n = 2, 3, \ldots$. The atom quickly returns to the ground state by emitting its excess energy in the form of light.

According to the quantum theory, light of a given frequency f consists of discrete particles called photons, each having energy $E = hf$, where h is *Planck's constant*. In addition, light obeys the usual wave relation $c = f\lambda$ where c is the speed of light, and λ is the wavelength so that for a photon $E = hc/\lambda$. The wavelength of the photon emitted in an atomic transition from the state n to the ground state is thus obtained by equating the energy of the photon to the energy given up by the atom. In the case of hydrogen this yields

$$E = \frac{hc}{\lambda_{n1}} = \Delta E_{n1} = E_n - E_1 = (13.6 \text{ eV})\left(1 - \frac{1}{n^2}\right)$$

or

$$\frac{1}{\lambda_{n1}} = \frac{\Delta E_{n1}}{hc} = (1.097 \times 10^{-2} \text{ nm}^{-1})\left(1 - \frac{1}{n^2}\right) \qquad 2$$

Of course, the atom need not return directly to the ground state, but can first make a transition to an intermediate state of energy E_n, and then make another transition to the ground state. The general relation for the energy emitted in an atomic transition from state n to state n' is therefore given by

$$\Delta E_{nn'} = E_n - E_{n'} = (13.6 \text{ eV})\left(\frac{1}{n'^2} - \frac{1}{n^2}\right) \qquad 3$$

The wavelength of the light emitted in such a transition is given by:

$$\frac{1}{\lambda_{nn'}} = \frac{\Delta E_{nn'}}{hc} = (1.097 \times 10^{-2} \text{ nm}^{-1})\left(\frac{1}{n'^2} - \frac{1}{n^2}\right) \qquad 4$$

In this experiment, the visible spectrum of hydrogen as well as the spectra of other atoms will be studied with an instrument called the *spectroscope*.

REFERENCE Cromer: *Physics for the Life Sciences*, Sec. 19.1 and 19.2.

Part I   Adjustment and Calibration of the Spectroscope

PROCEDURE

The essential components of the spectroscope are illustrated in Fig. 22.1. They consist of the collimator, prism, and telescope. The collimator is a tube with a slit at one end and a lens at the other. The distance between the slit and the lens is equal to the focal length of the lens. Light that enters the slit is rendered parallel by the lens. This light is incident upon one face of the prism where it is refracted and, because of the ability of a prism to disperse light into its component wavelengths, a spectrum of the incident light is produced. The objective lens of the telescope forms in its focal plane an image of the spectrum. This image is viewed through the magnifying eyepiece of the telescope.

The collimator and the prism remain fixed with respect to each other, while the telescope can be rotated through an arc sufficient to encompass the entire visible spectrum. In the third and shortest arm of the spectroscope there is a scale, the image of which comes into the field of view by reflection from one of the prism's faces. If this image is not visible, illuminate the scale by reflecting light on it from white paper, or by holding a small lamp at such a distance from the tube that the scale is just visible. Since the telescope is focused for parallel light, observations can be made most comfortably if the eye is relaxed, as in viewing a distant object.

Turn on the sodium-vapor lamp and place it so that the slit of the spectroscope is illuminated. Adjust the eyepiece of the telescope so that a sharp image of the yellow sodium line is seen. If necessary, adjust the length of the short tube to obtain a sharp image of the scale. The mark 3 on the scale should coincide with the sodium line. Now, move the eye from side to side and continue adjusting the short tube until there is no relative motion between the image of the sodium line and the image of the scale. If other adjustments are necessary, consult your instructor. Leave the sodium lamp on during the entire experiment for calibration checks. In addition to sodium, the lamp contains a small amount of argon, which may produce faint spectral lines of its own. Sketch these lines on the report sheet using colored pencils. Note the scale values.

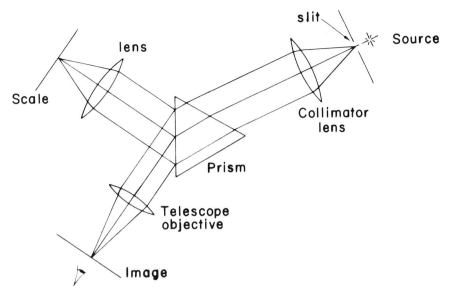

FIGURE 22.1   Schematic diagram of a spectroscope.

In all the observations that follow, it is necessary that the prism remain in the same position. This can be checked at any time with the sodium lamp by observing the coincidence of the yellow line and the mark 3 on the scale. The divisions on the scale are purely arbitrary. Before the instrument can be used for wavelength measurements the scale must be calibrated. This is done by using the known wavelengths in the visible spectrum of helium. Place the helium discharge tube securely in the spring-clip contacts on the transformer box. For safety's sake, keep your hands off the secondary terminals of the transformer and the connecting wires. The potential there and at the ends of the helium tube is nearly 10,000 V. Turn on the apparatus and point the slit of the spectroscope toward the capillary section of the helium discharge tube, adjusting the position of the instrument until the spectrum is clear and sharp. Read the positions of the nine helium lines to the nearest half division, and record them. Be sure to record the values for these lines in the order given on the report sheet (red to violet).

ANALYSIS

On the report sheet, draw a sketch of the helium spectrum with colored pencils. Label the proper wavelengths carefully.

Plot your observations of the scale reading of the spectroscope against the stated wavelength of the observed line. On your graph, denote each point with a dot and use a French curve to draw a very fine, smooth curve through your experimental points. This curve is the *calibration curve* for the spectroscope scale in terms of wavelength. For instance, if a certain spectral line falls on the scale at 4.83, its wavelength can be found from the calibration curve.

The measure of the spectroscope's ability to spread the spectrum out is called *dispersion*. For a given spectroscope, the dispersion may be conveniently expressed as the number of scale divisions per nanometer difference in the wavelength. This is equivalent to the *slope* of the calibration curve, which has just been plotted. Since this curve is not linear, the dispersion will vary from one part of the spectrum to another. A small portion of the curve, however, will be substantially linear and the slope in this region may be taken as $\Delta s/\Delta \lambda$, where Δs is the difference in scale reading for a difference in wavelength $\Delta\lambda$. Find Δs corresponding to $\Delta\lambda = 20$ nm for several region of the curve. Plot the dispersion $D = \Delta s/\Delta\lambda$ against the wavelength λ.

Part II   The Hydrogen Spectrum

PROCEDURE

Turn on the hydrogen discharge tube and observe the prominen lines that correspond to the first three lines of the Balmer series (H_α – red; H_β – blue; H_γ – violet). Find the wavelengths of these lines from the calibration curve and record the values.

ANALYSIS

Draw a sketch of the hydrogen spectrum with the colored pencils provided. This series of lines is the result of transitions from the higher levels in the hydrogen atom to the first excited level ($n' = 2$). Using Eq. 4 compute the wavelength of the transition from the third level ($n = 3$) down to the $n' = 2$ level. Compare this value with your measurements. Draw a diagram of the first five hydrogen-atom energy levels using Eq. 1 and indicate the Balmer series transitions.

Part III   Absorption Spectra

REMARK   This part of the experiment requires that the sun be shining. It may be necessary to rearrange the parts of the experiment so that this section can be performed.

PROCEDURE

Observe the spectrum produced by the light from an incadescent lamp. Then place the colored filters in front of the lamp, one by one, and note the difference.

Place the spectroscope near a window and observe the spectrum of sunlight. It may be necessary to put a small block under one leg of the instrument so that the slit will point toward the sky. Observe carefully the *dark* lines (Fraunhofer absorption lines) in the sun's spectrum and record the scale values. The dark lines are caused by gases in the outer mantle of the sun that absorb the light emitted from its surface. An atom can absorb light only at the same wavelengths at which it can emit light, because the electrons can make transitions only between the allowed energy levels. The dark lines, therefore, provide a "negative" picture of the emisstion spectra of the atoms present in the mantle of the sun.

The number of absorption lines will depend on the spectroscope used and on the brightness of the sky. If a relatively prominent line is not seen at the mark 3 on the scale, the prism may be out of adjustment. Check the alignment with the sodium lamp. Make a sketch of all your observations.

ANALYSIS

Explain briefly the effect of the filters on the incadescent lamp spectrum. From the calibration curve find the wavelengths of the more prominent Fraunhofer lines which you observed. Where possible, identify the elements corresponding to the lines in your table.

Name_____

Date_____

REPORT SHEET
EXPERIMENT 22   ATOMIC SPECTRA

Part I   Adjustment and Calibration of the Spectroscope

DATA

Helium Spectrum

Wavelength (nm)	Color	Scale reading
706.5		
667.8		
587.6		
504.7		
501.5		
492.2		
471.3		
447.1		
438.8		

ANALYSIS

Make a sketch of the helium spectrum with colored pencils.

Scale reading ⟶

Plot the data on the graph below (scale reading against wavelength). Draw in the calibration curve.

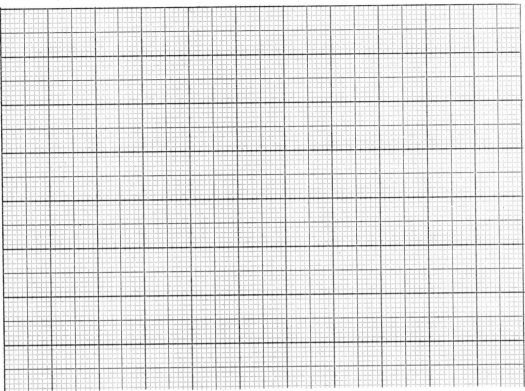

Calculate the dispersion, $D = \dfrac{\Delta s}{20 \text{ nm}}$, for three regions.

Region	D
~7000 nm	
~5500 nm	
~4500 nm	

Plot the dispersion D against wavelength.

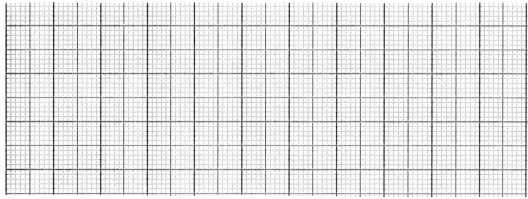

QUESTIONS

1. When there is no relative motion between the slit image and the image of the scale as the eye is moved from side to side, what do you know about the relative positions of these two images?
2. Suppose two spectral lines were separated by 1.0 nm in wavelength. Would you stand a better chance of separating their images if they were red or violet?

Part II   The Hydrogen Spectrum

DATA

Hydrogen wavelengths:

Line	Scale reading	Wavelength (nm)

ANALYSIS

Make a sketch of the hydrogen spectrum with the colored pencils:

_____

Wavelength of the $(n = 3)$ to $(n' = 2)$ transition:

$$\frac{1}{\lambda_{23}} = (1.097 \times 10^{-2})(\frac{1}{4} - \frac{1}{9}) = \underline{\hspace{2cm}} \text{ nm}^{-1}$$

$$\lambda_{23} = \underline{\hspace{2cm}} \text{ nm}$$

Draw a diagram of the first five hydrogen-atom energy levels and indicate on it the Balmer series transitions.

QUESTIONS

1. What is the value of the *smallest* wavelength line in the Balmer series?

2 Make a sketch of the entire Balmer series, working from Eq. 4, and identify those lines which you observed.

Part III Absorption Spectra

DATA

Describe the spectrum of incandescent lamp with and without filters.

Fraunhofer absorption lines:

Scale reading	Wavelength (nm)	Atom producing line

ANALYSIS

Explain briefly the effect of the filters on the spectrum of the incandescent lamp.

Find the wavelengths of the observed Fraunhofer lines from the calebration curve.

Identify the observed Fraunhofer lines with known spectral lines. Consult a chart of atomic spectra if necessary.

QUESTIONS

1 Explain how astronomers can determine the chemical composition of stars billions of miles away.
2 List some applications of spectroscopy to biology and chemistry.

EXPERIMENT 23   PHOTOELECTRIC EFFECT

GOALS

1 To demonstrate that the maximum energy of the electrons released in a photocell is proportional to the frequency of the incident light.

2 To measure Planck's constant.

EQUIPMENT

Photocell unit (Thornton PVB)
Amplifier power supply (Thornton APS)
Voltmeter
Ammeter (100 μA)
High-intensity lamp
Filters

INTRODUCTION

When electromagnetic radiation of frequency f is incident on a metallic surface, electrons will be ejected from the surface if f is greater than a specific *critical frequency* f_0 that is characteristic of the metal. The kinetic energies of the ejected electrons are distributed between zero and a maximum energy K_{max}, which is found experimentally to be related to f and f_0 by

$$K_{max} = h(f - f_0) = hf - E_0 \qquad 1$$

where h is a universal constant of nature, called *Planck's constant*, and $E_0 = hf_0$ is a constant characteristic of the metal.

The most significant aspect of the photoelectric effect is that K_{max} depends not on the intensity of the incident light, but only on the frequency. An increase in the intensity of the light increases the number of ejected electrons but not their maximum energy. This is contrary to what is expected on the basis of the wave nature of light, and it can be explained only by assuming that light has a corpuscular nature as well.

Einstein explained the photoelectric effect by assuming that light is composed of discrete corpuscles, called *photons*, each of which carries a definite energy $E = hf$. When a photon is absorbed on the surface of a metal, the entire energy E is transferred to a single electron. For any given metal, there is a minimum energy E_0 that an electron must expend in order to escape, so that an electron that acquires the energy $E = hf$ can escape from the metal with a kinetic energy of at most $K_{max} = E - E_0 = hf - E_0$. An electron can escape with a kinetic energy less than K_{max}, because an individual electron may expend more than the minimum energy E_0 in escaping, but an electron cannot leave the metal with a kinetic energy greater than K_{max}. As the intensity of the incident light is increased, the number of photons incident on the metal is increased, but not the energy of each photon. Hence, the number of ejected electrons is increased, but not their maximum kinetic energy.

In this experiment you will use a photocell to study the photoelectric effect and to measure Planck's constant. Figure 23.1 shows a schematic diagram of the equipment. The photocell consists of a metal plate and a collecting wire enclosed in an evacuated glass tube. Light of frequency f, incident on the plate, ejects electrons with energies between zero and K_{max}. Those electrons that reach the wire con-

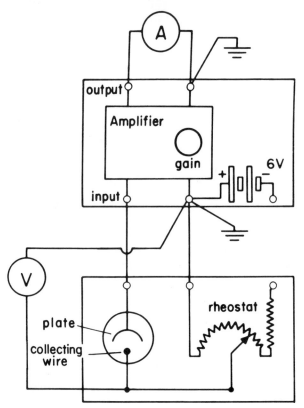

FIGURE 23.1 Schematic diagram of the photoelectric-effect experiment.

stitute a minute current in the external circuit, which can be amplified and read on a sensitive ammeter.

The collecting wire is maintained at a negative potential $-V$ with respect to the plate, so that only electrons with energies greater than eV can reach the wire. As the potential $-V$ is made more negative, the current in the photocell decreases, because fewer electrons have enough energy to reach the wire. At the stopping potential $-V_0 = -K_{max}/e$, none of the ejected electrons have enough energy to reach the wire, so the current is zero. Substituting the relation $K_{max} = V_0 e$ into Eq. 1, we get

$$V_0 = \frac{hf}{e} - \frac{E_0}{e} \qquad 2$$

In this experiment, the quantity h/e is found by measuring the stopping potential V_0 for light of several different frequencies.

REMARK This experiment is the counterpart of Experiment 14, in which the experiments that demonstrate the wave nature of light are performed.

REFERENCE Cromer: *Physics for the Life Sciences*, Sec. 19.1.

PROCEDURE

The photocell unit consists of a photocell and a rheostat to control the potential of the collecting wire. The rheostat is a wire resistor, which has a contact that can be moved by turning a knob. A fixed potential is maintained between the ends of the resistor, and a variable fraction of this potential is applied to the collecting wire of the photocell by a connection from the moveable contact. The potential across the photocell is measured by connecting a voltmeter between the moveable contact and the grounded end of the resistor (Fig. 23.1). The current in the photocell goes into an amplifier, and the output is measured by an ammeter.

REMARK The voltmeter cannot be placed directly across the photocell because too much current would flow into the amplifier through the voltmeter. However, since the input current to the amplifier is so small, there is only a minute potential difference between the ground and the input terminal of the amplifier. Consequently, the potential difference between the collecting wire and the plate is essentially the same as the potential difference between the collecting wire and the ground.

Connect the photocell unit as shown in Fig. 23.1. Adjust the rheostat so that the collecting wire is at about -2 V. Set the ammeter on the 100-μA scale, and turn up the gain of the amplifier to its maximum value. Adjust the offset control so that the ammeter needle is near zero. This is a delicate adjustment, but the exact position of the needle is not important.

Turn on an incandescent lamp, and place the bulb a few inches from the photocell. As long as the potential of the collecting wire is sufficiently negative, the ammeter deflection is not affected by the intensity of the light. Decrease the negative potential of the collecting wire until the ammeter has a substantial deflection. Some of the electrons ejected from the plate are now reaching the wire, and the number of electrons is proportional to the intensity of the light. Demonstrate this by noting how the ammeter deflection depends on the distance between the lamp and the photocell. This shows qualitatively that the number of ejected electrons depends on the intensity of the light, but that the energy of the electrons does not.

To make a more quantitative study, put a colored filter over the window of the photocell. Figure 23.2 shows the transmission properties of three typical filters. The important point to notice is that for each filter there is a shortest wavelength λ which it can transmit. This means that each filter transmits light with a maximum frequency $f = c/\lambda$ to the photocell, and so the maximum energy K_{max} of the electrons

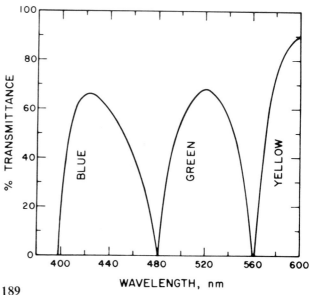

FIGURE 23.2   Transmittance curves of colored filters.

ejected from the plate of the photocell is given by Eq. 1, where f is the cutoff frequency of the filter.

Determine the stopping potential V_0 for several different filters by adjusting the rheostat until you obtain the smallest negative potential $-V_0$ for which the ammeter needle shows no deflection when an opaque card is alternately inserted and removed from between the lamp and the photocell. Each partner should make two independent determinations of V_0 for each filter.

ANALYSIS

Average the four determinations of the stopping potential made for each filter. Plot the potential V_0 against the cutoff frequency f of the filter. Draw a straight line through the points, and calculate the value of Planck's constant h from the slope of the line and the charge of the electron ($e = 1.60 \times 10^{-19}$ C). Compare your value of h with the accepted value ($h = 6.62 \times 10^{-34}$ J-s). In this experiment, a value of h within 50% of the accepted value is satisfactory.

Name_____

Date_____

REPORT SHEET
EXPERIMENT 23   PHOTOELECTRIC EFFECT

DATA

Filter color	Cut-off wavelength λ (m)	Cut-off frequency, f (Hz)	Stopping potential, V_0				
			Trial 1	Trial 2	Trial 3	Trial 4	Average

ANALYSIS

Plot the average value of V_0 against f. Determine h from the slope of the straight line drawn through the data.

Slope, s = _____

h = _____

QUESTIONS

1 Why is the photoelectric effect contrary to what is expected on the basis of the wave nature of light?

2 What is the minimum energy (in electron-volts) required to remove an electron from the plate of the photocell used in this experiment?

3 What is the smallest energy of light of wavelength 580 nm that can be absorbed by a rod cell in the retina of the eye? Rod cells can respond to a light stimulus this small.

EXPERIMENT 24   NUCLEAR RADIATION

GOALS

1 To learn to measure nuclear radiation with a Geiger tube.

2 To demonstrate that the intensity of nuclear radiation from a point source obeys the $1/r^2$ law.

3 To measure the half-layer value of gamma rays in lead and aluminum.

EQUIPMENT

Geiger tube and counter   Gamma-ray source (e.g., ^{137}Cs)
Timer   Lead and aluminum absorbers

INTRODUCTION

Radioactive nuclei usually decay by the emission of either alpha particles or beta rays. An alpha particle is a helium nucleus ($^{4}_{2}$He), so that its emission results in a daughter nucleus that has two less protons and two less neutrons than the parent nucleus. For example, $^{238}_{92}$U decays to $^{234}_{90}$Th by the emission of an alpha particle. The alpha particle emitted by a nucleus can travel only a few centimeters in air, and much less in solid matter, before it is brought to rest by a series of collisions with atomic electrons. Because of their short range, alpha particles are difficult to study in an elementary physics laboratory.

A beta ray is an electron that is emitted when a neutron inside a nucleus spontaneously changes into a proton. Consequently, the emission of a beta ray results in a daughter nucleus that has one more proton and one less neutron than the parent nucleus. For example, $^{137}_{55}$Cs decays to $^{137}_{56}$Ba by the emission of a beta ray. Although beta rays travel much farther than alpha particles, their range is also too short for convenient study. Low-energy beta rays (<1 MeV) are stopped by less than 5 mm of plastic and less than 1 mm of lead.

Gamma rays, which are very short wavelength radiation similar to x-rays, often accompany alpha and beta decay. For instance, only 4% of ^{137}Cs decays directly to the ground state of ^{137}Ba. The other 96% decays to an excited state of ^{137}Ba, which then decays to the ground state by the emission of a high-energy photon (gamma ray), just as an excited atom decays to the atomic ground state by the emission of a low-energy photon (light quantum). In this experiment you will study the properties of the gamma rays that accompany nuclear decay.

Gamma rays lose all their energy in a single interaction, unlike charged particles, which lose their energy gradually in a series of collisions. For instance, when a gamma-ray photon interacts with an atomic electron, all of the photon's energy is transferred to the electron, which consequently is ejected from the atom with a large kinetic energy. This process results in the destruction of the photon and the production of a high-energy electron, which behaves as a beta ray.

The distance a gamma ray will travel before it interacts with an electron cannot be predicted. All that can be predicted is the distance in which a gamma ray has a 50% chance of interacting. This distance is called the *half-value layer* λ, and it depends on the energy of the gamma ray and the density of the material through which it is traveling. In this experiment you will measure the half-value layer of gamma rays in lead and aluminum.

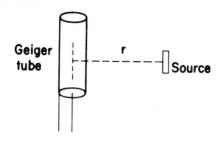

FIGURE 24.1   Schematic diagram of a Geiger tube.

Nuclear radiation is detected by a device called a *Geiger tube*, which is a metallic gas discharge tube with a wire running through its center (Fig. 24.1). The wire is maintained at a large positive potential (about +500 V) relative to the wall of the tube, and the tube contains a gas at low pressure (about 0.1 atm). When a single charged particle passes through the gas in the tube, it ionizes one or more atoms in the gas, releasing several electrons. These electrons are then accelerated toward the wire, but before they reach it they usually ionize other atoms in their path, thereby releasing more electrons. These additional electrons in turn ionize still more atoms as they accelerate toward the wire. A chain reaction (or *avalanche*) develops, which rapidly ionizes most of the gas in the tube. As a consequence, the gas becomes electrically conducting and there is a momentary surge of current in the tube. This current pulse goes to an amplifier, and the output of the amplifier operates a counter, which records the discharge produced each time a charged particle passes through the tube.

Geiger tubes do not detect gamma rays directly. However, a small fraction (about 2%) of the gamma rays incident on the tube knock electrons out of the atoms in the tube, and these electrons discharge the tube. Thus, a Geiger tube detects a definite proportion of the gamma rays that are incident on it.

REFERENCE   Cromer: *Physics for the Life Sciences*, Sec. 20.1, 20.2, and 20.4.

Part I $1/r^2$ Law

PROCEDURE

Gamma rays are emitted in all directions from a nuclear source. These rays, like all other radiation, travel in straight lines, so that the intensity of the radiation decreases as the square of the distance from the source (Cromer: *Physics for the Life Sciences*, Sec. 13.3). We shall study the dependence of the intensity on distance as an introduction to the techniques of radiation detection.

FIGURE 24.2   Arrangement of source and Geiger tube for the $1/r^2$-law experiment.

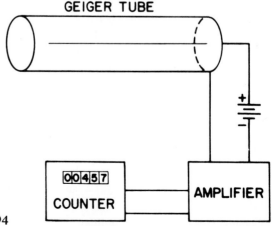

Your instructor will show you how to connect the Geiger tube to the counter, and how to start and reset the counter. When you understand the operation of the counter, mount the Geiger tube in a fixed horizontal position, and move the radiation source at least 6 ft from the Geiger tube. Turn on the counter and count the background radiation for 5 min. There is always a small background of radiation present due to the cosmic rays that reach the earth from outer space. This background must be subtracted from all your subsequent measurements before they can be analyzed properly. Record the number of counts and the exact counting time.

Place the radiation source along the side of the Geiger tube, opposite the middle of the sensitive region (Fig. 24.2). Measure the distance r from the midline of the tube to the center of the source. Take a series of six 2-min counts with r between 4 cm and 30 cm. (If the tube has a thin window on its side, turn the window away from the source. Gamma rays can pass through the wall of the tube, but alpha particles and beta rays are stopped by it.)

ANALYSIS

Calculate the background counting rate R_B in counts per minute (cpm) by dividing the number of background counts by the counting time. Determine the corrected counting rate R for each of your measurements by subtracting the background rate R_B from the uncorrected rate R'.

Plot R against $1/r^2$, where r is the distance between the source and the detector. The points should lie on a straight line that passes through the origin.

Part II   Gamma Ray Absorption

PROCEDURE

Mount the Geiger tube in a vertical position about 5 cm above a radiation source. Remove the source and measure the background.

Replace the source and count for 1 min. Then take 1-min counts with successive sheets of lead placed over the source. Continue adding layers until the counting rate is less than 1/10 the rate with no lead sheets.

Repeat using aluminum sheets instead of lead. Measure the thicknesses of the lead and aluminum sheets.

ANALYSIS

Calculate the corrected counting rate for each measurement.

The gamma ray intensity decreases exponentially with the thickness of the absorber. This is expressed mathematically by the formula

$$I = I_0 2^{-d/\lambda}$$

where I_0 is the intensity incident on the absorber, I is the intensity after passing through a thickness d of absorber, and λ is the half-value layer. Taking the logarithm of both sides of this equation we get

$$\begin{aligned}\log I &= \log(I_0 2^{-d/\lambda}) \\ &= \log I_0 - (d/\lambda)\log 2 \\ &= A - Bd\end{aligned}$$

where $A = \log I_0$ and $B = \lambda^{-1} \log 2$ are constants. This equation shows that $\log I$ is a linear function of the thickness of the absorber. Since the counting rate R is proportional to I, a plot of $\log R$ against d

should give a straight line.

Plot R against d on semilogarithmic paper for both lead and aluminum absorbers. Draw straight lines through your data and determine the half-layer value for both lead and aluminum from the thickness at which R has decreased by a factor of 2.

Your instructor will tell you the energy of the gamma rays emitted from your source. Record this information, because λ depends on the gamma-ray energy.

Name_____

Date_____

REPORT SHEET
EXPERIMENT 24   NUCLEAR RADIATION

Part I $1/r^2$ Law

DATA

Background: Number of counts, N = _____

Counting time, t = _____ min

Counting rate, $R_B = N/t$ = _____ cpm

Distance r (cm)	No. of counts N (counts)	Time t (min)	Uncorrected counting rate R' (cpm)	Corrected counting rate R (cpm)	$\frac{1}{r^2}$ (cm^{-2})

ANALYSIS

Determine the corrected counting rate R, and complete the data table. Plot R against $1/r^2$.

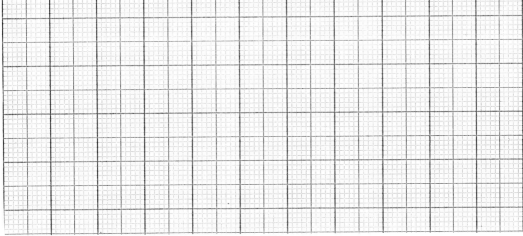

QUESTIONS

1. Why will your data not follow a $1/r^2$ law when r is very small?
2. Explain why the intensity of any radiation that emanates uniformly in all directions from a point source must obey a $1/r^2$ law.
3. A plot of $\log R$ against $\log r$ also gives a straight line. (*a*) What is the slope of this line? (*b*) Why is such a log-log plot better than an R-vs-$1/r^2$ plot for verifying a $1/r^2$ law?

Part II   Gamma Ray Absorption

DATA

Background: Number of counts, $N =$ _____

Counting time, $t =$ _____ min

Counting rate, $R_B = N/t =$ _____ cpm

Absorber	Number of sheets n	Thickness d $(n\Delta)$	No. of counts N	Time t (min)	Uncorrected counting rate R' (cpm)	Corrected counting rate R (cpm)
Lead: thickness of each sheet, $\Delta =$ _____						
Aluminum: thickness of each sheet, $\Delta =$ _____						

ANALYSIS

Determine the corrected counting rate R and complete the data table; on the semilogrithmic paper, plot R against d for each absorber. Determine the half-value layer λ of each absorber.

$\lambda_{Pb} =$ _____

$\lambda_{Al} =$ _____

Energy of the gamma ray = _____

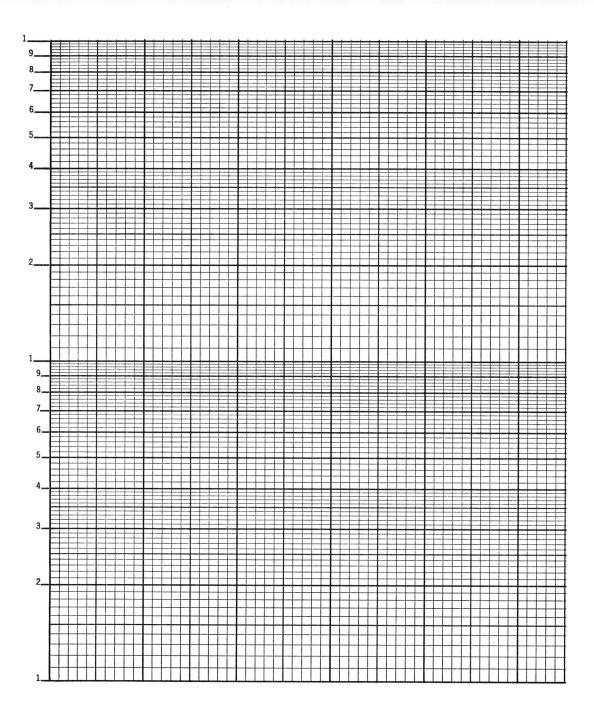

QUESTIONS

1 What thickness of lead is required to reduce the intensity of gamma rays from your source by a factor of 10?

2 Draw a semilogrithmic plot of R against d for 1-Mev gamma rays, which have a half-layer value in lead of 0.89 cm. What thickness of lead is required to reduce the intensity of 1-Mev gamma rays by a factor of 20?

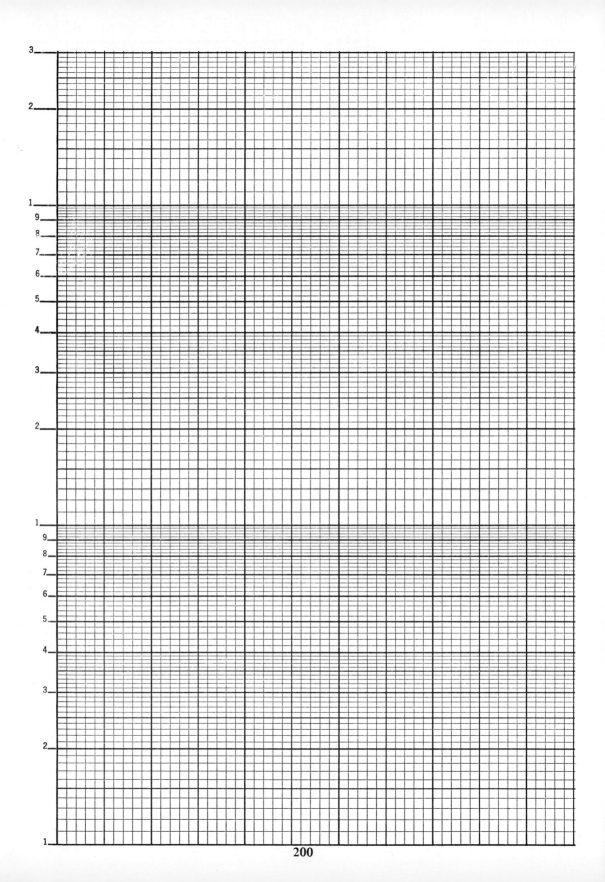

EXPERIMENT 25   NUCLEAR DECAY

GOALS

1 To demonstrate that radioactive decay is a random process.

2 To measure the half life of $^{137}Ba^m$.

EQUIPMENT

Geiger tube and counter
Timer
Radiation source
$^{137}Cs/^{137}Ba$ Minigenerator

INTRODUCTION

Since the decay of a radioactive nucleus is a spontaneous random event, the time at which an individual nucleus will decay cannot be predicted. All that can be predicted is the probability that a nucleus will decay within a given period. This is usually expressed by giving the period τ within which a nucleus has a 50% chance of decaying. This period, called the *half-life*, is characteristic of each radioactive isotope. In this experiment you will measure the half-life of an isotope.

Suppose that at time $t = 0$ a sample contains N nuclei of a radioactive isotope whose half-life is τ. Then, at time $t = \tau$ half of these nuclei will have decayed, and ½N nuclei will be left undecayed. At time $t = 2\tau$, half of the ½N nuclei will have decayed, and ¼N will be left undecayed. At time $t = 3\tau$, only 1/8 N nuclei remain undecayed.

Although the number N of nuclei in a sample cannot be easily measured, the rate of decay R can be. At any instant of time the decay rate is

$$R = \frac{0.693}{\tau} N \qquad \qquad 1$$

This means that R decreases as N decreases, so the half-life of an isotope can be determined by measuring the decrease in the decay rate with time. That is, the decay rate at time $t = \tau$ will be half the decay rate at time $t = 0$, because there are only ½N nuclei left undecayed at time $t = \tau$.

Because of the random nature of the decay process, the determination of R is subject to statistical uncertainties. Suppose, for instance, that an average counting rate per minute from a radioactive sample is determined by counting for 1 hr and dividing the total number of counts by 60. The number of counts recorded in any one 1-min interval will differ from this average, or "true", counting rate because in any one 1-min interval there may, by chance, be a few more or less decays than average. Thus, a single 1-min measurement may differ considerably from the "true" counting rate.

It is shown in statistics that if $\bar{n}$ is the average number of counts expected in a given time interval, then 68% of the time the actual number n of counts measured in this interval will be between $\bar{n} - \sqrt{\bar{n}}$ and $\bar{n} + \sqrt{\bar{n}}$, and 32% of the time n will differ from $\bar{n}$ by more than $\sqrt{\bar{n}}$. If n is the number of counts recorded in a single measurement, the true average $\bar{n}$ will be between $n - \sqrt{n}$ and $n + \sqrt{n}$ about 68% of the time. Thus, from a single measurement the average is estimated to be

$$\bar{n} = n \pm \sqrt{n} \qquad \qquad 2$$

where $\sqrt{n}$ is called the *standard deviation* of the measurement. The *relative error* of the measurement is $\sqrt{n}/n = 1/\sqrt{n}$, which decreases as n increases.

We shall study the statistical nature of nuclear decay in Part I of this experiment, and measure the half life of $^{137}Ba^m$ in Part II.

REFERENCE Cromer: *Physics for the Life Sciences*, Sec. 20.2, Appendix VII, and Experiment 24.

Part I   Counting Statistics

PROCEDURE

Place a radioactive source at a distance from a Geiger tube such that the counting rate is about 50 cpm. Start the counter at time $t = 0$ and record the accumulated number of counts at 10-s intervals for 10 min. Be careful to take each reading exactly at the 10-s mark.

ANALYSIS

Determine the number of counts n_l in the lth 10-s interval by subtracting the accumulated counts at time t_l from the accumulated counts at time $t_{l+1} = t_l + 10$ s. Find the average number of counts $\bar{n}$ in a 10-s interval by dividing the total number of accumulated counts by the number of 10-s intervals during which they were obtained. Since you are averaging a relatively large number of 10-s intervals, $\bar{n}$ is a good approximation to the "true" average number of counts.

Make a histogram, or bar graph, of the number of occurrences of each value of n. An example is shown in Fig. 25.1. The bar at $n = 6$ is drawn 4 units high to indicate that 6 counts were recorded in 4 different 10-s intervals. The histogram clearly shows the random nature of nuclear decay.

Draw a vertical line on your histogram to represent the average number of counts $\bar{n}$. The standard deviation σ of a distribution is defined as the number such that 68% of the distribution is between $\bar{n} - \sigma$ and $\bar{n} + \sigma$. Statistics shows that for random events, such as nuclear decay, σ is equal to $\sqrt{\bar{n}}$. Draw vertical lines on your histogram to represent $\bar{n} - \sqrt{\bar{n}}$ and $\bar{n} + \sqrt{\bar{n}}$, and determine the fraction of your distribution that lies between these lines. In the example shown in Fig. 25.1, this fraction is 66%, which is close to the expected value of 68%. Thus, most individual counts lie between $\bar{n} - \sqrt{\bar{n}}$ and $\bar{n} + \sqrt{\bar{n}}$.

Part II   Half Life of $^{137}Ba^m$

PROCEDURE

The Union Carbide Minigenerator^R provides a convenient source of a short-lived radioactive isotope that is suitable for laboratory work. It operates on the same principle as generators used to provide short-lived isotopes (such as $^{99}Tc^m$) in hospitals. A long-lived parent isotope is attached to a substrate for which it has great chemical affinity. The parent decays to a short-lived daughter that has less affinity for the substrate, and so can be eluded by passing a suitable solvent through the substrate.

The $^{137}Cs/^{137}Ba$ generator consists of ^{137}Cs, which beta decays with a half life of 30 years to stable ^{137}Ba. However, only 4% of these decays go directly to the ^{137}Ba ground state. The other 96% first decay to $^{137}Ba^m$, an excited state of ^{137}Ba, which then decays to the ^{137}Ba ground state by the emission of a gamma ray. This particular excited state has an exceptionally long half-life (2.5 min), so it can be treated as a separate radioactive nuclide. The superscript m indicates that it is metastable, or long-lived excited state of ^{137}Ba. A sample of $^{137}Ba^m$ is obtained by passing a dilute NaCl-HCl solution through the $^{137}Cs/^{137}Ba$ generator.

Mount your Geiger tube in a vertical position about 1 cm above a glass plate. Count the background for several minutes.

Have your instructor show you how to elude several drops of $^{137}Ba^m$ from the generator. Put these drops on the glass plate under the Geiger tube. Start the timer and count for 30 s. Then wait 30 s and again count for 30 s. Continue counting for 30 s and waiting for 30 s, until your counting rate is close to the background rate. Record the times t_1 and t_2 at which each counting interval began and ended, and the number n of counts recorded during the interval.

ANALYSIS

Calculate the corrected counting rate R for each measurement by subtracting the background rate.

The number N of $^{137}Ba^m$ nuclei in the sample at time t is given by

$$N = N_0 2^{-t/\tau} \qquad\qquad 2$$

where τ is the half life, and N_0 is the number of $^{137}Ba^m$ nuclei present at time $t = 0$. From Eqs. 1 and 2 we see that the rate of decay, and hence the observed counting rate, is given by

$$R = \frac{0.693 \, N_0}{\tau} 2^{-t/\tau} = R_0 2^{-t/\tau}$$

Taking the logarithm of both sides of this equation we get

$$\log R = \log R_0 - \left(\frac{t}{\tau}\right) \log 2$$
$$= A - Bt$$

where $A = \log R_0$ and $B = \tau^{-1} \log 2$ are constants. Thus, a plot of $\log R$ against t will give a straight line.

Plot R against t_1 on semilog arithmic paper. Draw a straight line through your data and determine the half life of $^{137}Ba^m$ from the time in which R decreases by a factor of 2.

FIGURE 25.1 Histogram of the number of occurrences of n counts in 60 10-s intervals when the average value of n is 9.

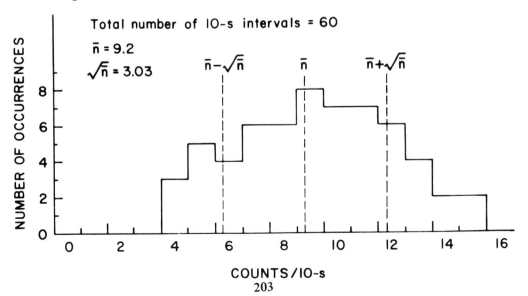

Name_____

Date_____

REPORT SHEET
EXPERIMENT 25   NUCLEAR DECAY

Part I   Counting Statistics

DATA

Time (s)	Accum. counts
0	0
10	
20	
30	
40	
50	
60	
70	
80	
90	
100	
110	
120	
130	
140	
150	

Time (s)	Accum. counts
160	
170	
180	
190	
200	
210	
220	
230	
240	
250	
260	
270	
280	
290	
300	

Time (s)	Accum. counts
310	
320	
330	
340	
350	
360	
370	
380	
390	
400	
410	
420	
430	
440	
450	

Time (s)	Accum. counts
460	
470	
480	
490	
500	
510	
520	
530	
540	
550	
560	
570	
580	
590	
600	

ANALYSIS

Determine the number of counts n_l in each 10-s interval.

l	n_l
1	
2	
3	
4	
5	
6	
7	
8	
9	
10	

l	n_l
11	
12	
13	
14	
15	
16	
17	
18	
19	
20	

l	n_l
21	
22	
23	
24	
25	
26	
27	
28	
29	
30	

l	n_l
31	
32	
33	
34	
35	
36	
37	
38	
39	
40	

l	n_l
41	
42	
43	
44	
45	
46	
47	
48	
49	
50	

l	n_l
51	
52	
53	
54	
55	
56	
57	
58	
59	
60	

Calculate the average number of counts in 60 10-s intervals.

$$\bar{n} = \frac{\text{total number of counts}}{60} = \underline{\hspace{3cm}}$$

$\sigma = \sqrt{\bar{n}} = \underline{\hspace{3cm}}$

Plot a histogram of the number of occurrences of each value of n. Draw vertical lines at $\bar{n}$, $\bar{n} - \sigma$, and $\bar{n} + \sigma$. Determine the fraction f of the distribution that lies between $\bar{n} - \sigma$ and $\bar{n} + \sigma$.

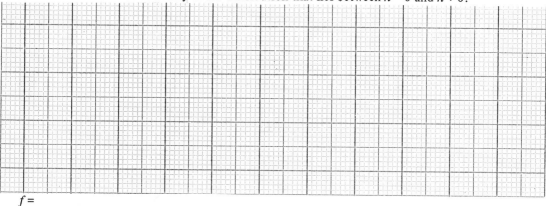

$f = \underline{\hspace{4cm}}$

QUESTIONS

1. A Geiger counter records 650 counts in 5.0 min. (*a*) What is the average counting rate per minute, and the standard error of the counting rate? (*b*) What is the relative error? (*c*) About how long would one have to count to get a relative error of 1%?

2. The average counting rate in a given situation is 486 cpm. What is the probability that in any given 10-s interval one will get less than 72 counts? Is this the same as the probability of getting less than $72 \times 6 = 432$ counts in 60 s?

Part II   Half-Life of $^{137}\text{Ba}^m$

DATA

Background: Number of counts = _____

Counting time, t = _____ min

Counting rate R_B = _____ cpm

Starting time t_1 (s)	Starting time t_2 (s)	No. of counts n	Uncorrected counting rate R' (cpm)	Corrected counting rate R (cpm)
0	30			
60	90			
120	150			
180	210			
240	270			
300	330			
360	390			
420	450			

ANALYSIS

Calculate the corrected counting rate for each 30-s interval. Plot R against t_1 on the semilogarithm paper. Draw a straight line through the points, and determine the half-life from the line.

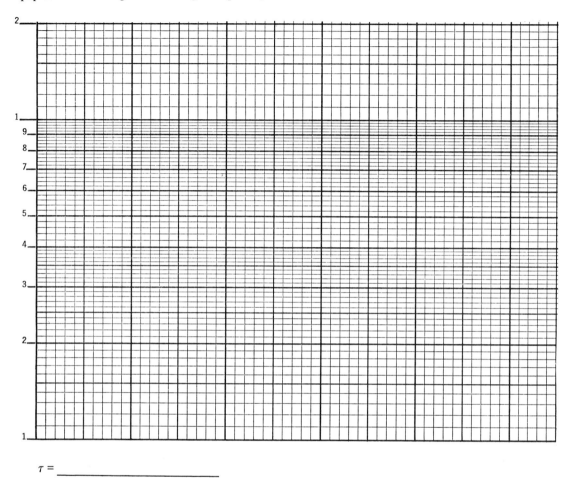

$T = $ _____

QUESTIONS

1. What effect does the randomness of nuclear decay have on the accuracy of this experiment? How can the accuracy be increased?
2. A radioactive substance gives a counting rate of 1250 cpm at time $t = 0$, and a counting rate of 475 cpm at time $t = 20$ min. What is the half-life of the substance?
3. The half-life of the parent nuclide (^{137}Cs) in the minigenerator is 30.2 years. If the current activity in a generator is 8.5 μC (1 μC = 3.6 × 10^4 decays per second), what will be its activity in 15 years?

TABLE OF TRIGONOMETRIC FUNCTIONS

Sine, cosine, and tangent of angles from 0 to 90°. For small angles ($\theta < 10°$), the cosine is approximately 1, and the sine and tangent are approximately equal. The value of the sine and tangent for angles less than 10° is given to good approximation by $\sin \theta = \tan \theta = 0.0174$.

Angle (deg)	Sine	Co-sine	Tan-gent	Angle (deg)	Sine	Co-sine	Tan-gent
0	0.000	1.000	0.000				
1	.017	1.000	.017	46	0.719	0.695	1.036
2	.035	.999	.035	47	.731	.682	1.072
3	.052	.999	.052	48	.743	.669	1.111
4	.070	.998	.070	49	.755	.656	1.150
5	.087	.996	.087	50	.766	.643	1.192
6	.105	.995	.105	51	.777	.629	1.235
7	.122	.993	.123	52	.788	.616	1.280
8	.139	.990	.141	53	.799	.602	1.327
9	.156	.988	.158	54	.809	.588	1.376
10	.174	.985	.176	55	.819	.574	1.428
11	.191	.982	.194	56	.829	.559	1.483
12	.208	.978	.213	57	.839	.545	1.540
13	.225	.974	.231	58	.848	.530	1.600
14	.242	.970	.249	59	.857	.515	1.664
15	.259	.966	.268	60	.866	.500	1.732
16	.276	.961	.287	61	.875	.485	1.804
17	.292	.956	.306	62	.883	.469	1.881
18	.309	.951	.325	63	.891	.454	1.963
19	.326	.946	.344	64	.899	.438	2.050
20	.342	.940	.364	65	.906	.423	2.145
21	.358	.934	.384	66	.914	.407	2.246
22	.375	.927	.404	67	.921	.391	2.356
23	.391	.921	.424	68	.927	.375	2.475
24	.407	.914	.445	69	.934	.358	2.605
25	.423	.906	.466	70	.940	.342	2.747
26	.438	.899	.488	71	.946	.326	2.904
27	.454	.891	.510	72	.951	.309	3.078
28	.469	.883	.532	73	.956	.292	3.271
29	.485	.875	.554	74	.961	.276	3.487
30	.500	.866	.577	75	.966	.259	3.732
31	.515	.857	.601	76	.970	.242	4.011
32	.530	.848	.625	77	.974	.225	4.331
33	.545	.839	.649	78	.978	.208	4.705
34	.559	.829	.675	79	.982	.191	5.145
35	.574	.819	.700	80	.985	.174	.5671
36	.588	.809	.727	81	.988	.156	6.314
37	.602	.799	.754	82	.990	.139	7.115
38	.616	.788	.781	83	.993	.122	8.144
39	.629	.777	.810	84	.995	.105	9.514
40	.643	.766	.839	85	.996	.087	11.43

41	.656	.755	.869	86	.998	.070	14.30
42	.669	.743	.900	87	.999	.052	19.08
43	.682	.731	.933	88	.999	.035	28.64
44	.695	.719	.966	89	1.000	.017	57.29
45	.707	.707	1.000	90	1.000	.000	∞

TABLE OF SQUARE ROOTS

To find the square root of a number less than 1 or greater than 100, write the number as a number between 1 and 100 times an even power of 10. Thus, if $n = 460,000$, write it as $n = 46 \times 10^4$, so that

$$\sqrt{n} = \sqrt{46} \times 10^2 = 678$$

and, if $n = 0.018$, write it as 1.8×10^{-2}, so that

$$\sqrt{n} = \sqrt{1.8} \times 10^{-1} = 0.134$$

n	$\sqrt{n}$	n	$\sqrt{n}$	n	$\sqrt{n}$	n	$\sqrt{n}$
1.00	1.00	3.80	1.95	10.0	3.16	38.0	6.16
1.10	1.05	4.00	2.00	11.0	3.32	40.0	6.32
1.20	1.10	4.20	2.05	12.0	3.46	42.0	6.48
1.30	1.14	4.40	2.10	13.0	3.61	44.0	6.63
1.40	1.18	4.60	2.14	14.0	3.74	46.0	6.78
1.50	1.22	4.80	2.19	15.0	3.87	48.0	6.93
1.60	1.26	5.00	2.24	16.0	4.00	50.0	7.07
1.80	1.34	5.50	2.35	18.0	4.24	55.0	7.42
2.00	1.41	6.00	2.45	20.0	4.47	60.0	7.75
2.20	1.48	6.50	2.55	22.0	4.69	65.0	8.06
2.40	1.55	7.00	2.65	24.0	4.90	70.0	8.37
2.60	1.61	7.50	2.74	26.0	5.10	75.0	8.66
2.80	1.67	8.00	2.83	28.0	5.29	80.0	8.94
3.00	1.73	8.50	2.92	30.0	5.48	85.0	9.22
3.20	1.79	9.00	3.00	32.0	5.66	90.0	9.49
3.40	1.84	9.50	3.08	34.0	5.83	95.0	9.75
3.60	1.90	10.00	3.16	36.0	6.00	100.0	10.00

INDEX OF EQUIPMENT

Numbers refer to the number of the experiment in which the equipment is used.

Absorbers, lead and aluminum, 24
Air track, 4, 5
Ammeter, 18, 19, 21, 22
Amplifier, 23
    differential, 20

Balance, triple-beam, 1, 3, 6, 13
Battery, 20
Beaker, 8, 10, 11, 12
Board and pins, 15
Bunsen burner, 10

Caliper, vernier, 1
Capillaries, glass, 8, 12
Carbon tetrachloride (Tetrachloromethane), 11
Clamps, stands, and rods, 2, 3, 5, 7, 8, 10, 11
Clamps for tubing, 8, 11
Color filters, 14, 22, 23
Compass, 21
Conducting board with double-contact probe, 17
Constant-level cups, 8
Cord, 2, 3, 4, 5, 13
Counter-timer, electronic, 5
Cylinder, graduated (100 ml), 1, 8
Cylinders, metal, 1

Detergent, 12
Diffraction grating, 14

Electric-field apparatus, 17
Electrocardiogram electrodes and paste, 20

Filters, color, 14, 22, 23
Flexible-tube manometer, 9
    with glass bulb, 10, 11
Force board, 2

Gamma-ray source, ^{137}Cs, 24, 25
    Minigenerator, 25
Geiger tube and counter, 24, 25
Graduated cylinder (100 ml), 1, 8
Ground-glass screen, 16

Helmholtz coils, 21

Ice, 10, 11
Image light, 16
Iron filings, 21

Jar magnet, 21

Lenses, 16
Light source, incandescent, 22, 23
Light bulb (10 V), 18

Magnets, 21
Manometer, mercury, 7
    flexible tube, 9
        with glass bulb, 10, 11
Mechanical equivalent of heat apparatus,
    Cavendish form, 6
Meterstick, 3, 14
Micrometer, 1
Minigenerator, ^{137}Cs/^{137}Ba, 25
Mirror, plane, 15

Optical bench, 14, 16
Oscillator, audio, 20
Oscilloscope, 20

Pencils, colored, 22
Photoelectric effect apparatus, 23
Photogates, 5
Pin board and pins, 15
Pivoted meterstick, 3
Power supply, 17, 18, 19, 21, 23
Projector, overhead, 21
Protractor, 2, 3, 15
Pulleys, 2, 4, 5, 13

Ray box, 15
Relay, mercury, 20
Resistors, 18, 19
Reversing switch, 21
Rheostat (22-ohm), 21
Rods, clamps, and stands, 2, 3, 5, 7, 8, 10, 11
Ruler, 1, 12, 15

Shot and metal can, 13
Slit slide, 14
Sodium lamp, 14, 22
Spark timer, 4
Spark-sensitive paper, 4
Speaker, 20
Spectra tubes with power supply,
    helium, 22

    hydrogen, 22
    mercury, 14
Spectrometer, prism, 22
Spring scales, 2, 3
Standing wave in air apparatus, 13
Stands, rods, and clamps, 2, 3, 5, 7, 8, 10, 11
Supports, inverted-v, 3
Surface-tension apparatus, 12
Syringes with platforms mounted on plunger, 7

T-connector, 11
Tetrachloromethane, CCl_4, 11
Thermometer, 6, 10, 11
Timer, 6, 8, 21, 24, 25

Transformer,
    audio, 20
    6.3 V ac, 20
Transparent board, 21
Tubing, rubber, 7, 8
Tuning forks, 13

Vibrator, 13
Voltmeter, 18, 19, 23
    VTVM, 17

Weights, 2, 3, 4, 5, 7
Wire leads, 18, 19

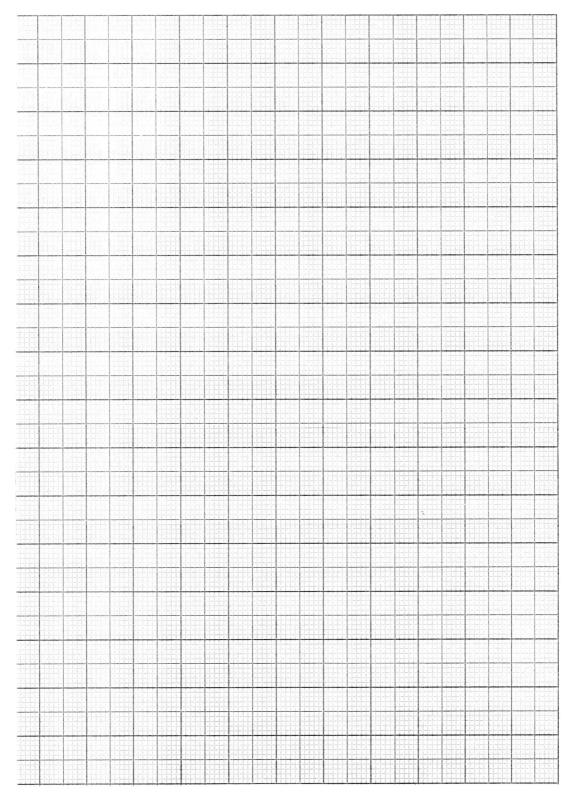